T0281705

HYDRODYNAMICS OF PUMPS

Hydrodynamics of Pumps is a reference for pump experts and a textbook for advanced students exploring pumps and pump design. This book is about the fluid dynamics of liquid turbomachines, particularly pumps. It focuses on special problems and design issues associated with the flow of liquid through a rotating machine. There are two characteristics of a liquid that lead to problems and cause a significantly different set of concerns from those in gas turbines. These are the potential for cavitation and the high density of liquids, which enhances the possibility of damaging, unsteady flows and forces. The book begins with an introduction to the subject, including cavitation, unsteady flows, and turbomachinery as well as basic pump design and performance principles. Chapter topics include flow features, cavitation parameters and inception, bubble dynamics, cavitation effects on pump performance, and unsteady flows and vibration in pumps – discussed in the three final chapters. The book is richly illustrated and includes many practical examples.

Christopher E. Brennen is Professor of Mechanical Engineering in the Faculty of Engineering and Applied Science at the California Institute of Technology. He has published more than 200 refereed articles and is especially well known for his research on cavitation and turbomachinery flows, as well as multiphase flows. He is the author of *Fundamentals of Multiphase Flows* and *Cavitation and Bubble Dynamics* and has edited several other works.

Hydrodynamics of Pumps

CHRISTOPHER E. BRENNEN

California Institute of Technology

CAMBRIDGE
UNIVERSITY PRESS

CAMBRIDGE UNIVERSITY PRESS
Cambridge, New York, Melbourne, Madrid, Cape Town,
Singapore, São Paulo, Delhi, Mexico City

Cambridge University Press
32 Avenue of the Americas, New York NY 10013-2473, USA

Published in the United States of America by Cambridge University Press, New York

www.cambridge.org
Information on this title: www.cambridge.org/9781107401495

First published 1994 by Oxford University Press
Cambridge University Press published 2011
First published 2011
First paperback edition 2013

A catalogue record for this publication is available from the British Library

Library of Congress Cataloguing in Publication Data
Brennen, Christopher E. (Christopher Earls), 1941–
  Hydrodynamics of pumps / Christopher Earls Brennen.
    p. cm.
  Includes bibliographical references and index.
  ISBN 978-1-107-00237-1 (hardback)
  1. Pumping machinery–Fluid dynamics. 2. Hydrodynamics. I. Title.
  TJ901.B74 2011
  621.2'52–dc22   2010043266

ISBN 978-1-107-00237-1 Hardback
ISBN 978-1-107-40149-5 Paperback

Contents

Preface

This book is intended as a combination of a reference for pump experts and a monograph for advanced students interested in some of the basic problems associated with pumps. It is dedicated to my friend and colleague Allan Acosta, with whom it has been my pleasure and privilege to work for many years.

But this book has other roots as well. It began as a series of notes prepared for a short course presented by Concepts NREC and presided over by another valued colleague, David Japikse. Another friend, Yoshi Tsujimoto, read early versions of the manuscript and made many valuable suggestions.

It was a privilege to have worked on turbomachinery problems with a group of talented students at the California Institute of Technology, including Sheung-Lip Ng, David Braisted, Javier Del Valle, Greg Hoffman, Curtis Meissner, Edmund Lo, Belgacem Jery, Dimitri Chamieh, Douglas Adkins, Norbert Arndt, Ronald Franz, Mike Karyeaclis, Rusty Miskovish, Abhijit Bhattacharyya, Adiel Guinzburg, and Joseph Sivo. I recognize the many contributions they made to this book.

In the first edition, I wrote that this work would not have been possible without the encouragement, love, and companionship of my beloved wife Doreen. Since then fate has taken her from me and I dedicate this edition to our daughters, Dana and Kathy, whose support has been invaluable to me.

Christopher E. Brennen
California Institute of Technology
January 2010

Nomenclature

Roman letters

a	Pipe radius
A	Cross-sectional area
A_{ijk}	Coefficients of pump dynamic characteristics
$[A]$	Rotordynamic force matrix
Ar	Cross-sectional area ratio
B	Breadth of passage or flow
$[B]$	Rotordynamic moment matrix
c	Chord of the blade or foil
c	Speed of sound
c	Rotordynamic coefficient: cross-coupled damping
c_b	Interblade spacing
c_{PL}	Specific heat of liquid
C	Compliance
C	Rotordynamic coefficient: direct damping
C_D	Drag coefficient
C_L	Lift coefficient
C_p	Coefficient of pressure
C_{pmin}	Minimum coefficient of pressure
d	Ratio of blade thickness to blade spacing
D	Impeller diameter or typical flow dimension
Df	Diffusion factor
D_T	Determinant of transfer matrix $[T]$
e	Specific internal energy
E	Energy flux
E	Young's modulus
f	Friction coefficient
F	Force
g	Acceleration due to gravity

g_s	Component of g in the s direction
h	Specific enthalpy
h	Blade tip spacing
h_p	Pitch of a helix
h^T	Total specific enthalpy
h^*	Piezometric head
H	Total head rise
$H(s,\theta,t)$	Clearance geometry
I	Acoustic impulse
I, J	Integers such that $\omega/\Omega = I/J$
I_P	Pump impedance
j	Square root of -1
k	Rotordynamic coefficient: cross-coupled stiffness
k_L	Thermal conductivity of the liquid
K	Rotordynamic coefficient: direct stiffness
K_G	Gas constant
ℓ	Pipe length or distance to measuring point
L	Lift
L	Inertance
L	Axial length
$\mathcal{L}$	Latent heat
m	Mass flow rate
m	Rotordynamic coefficient: cross-coupled added mass
m_G	Mass of gas in bubble
m_D	Constant related to the drag coefficient
m_L	Constant related to the lift coefficient
M	Moment
M	Mach number, u/c
M	Rotordynamic coefficient: direct added mass
n	Coordinate measured normal to a surface
N	Specific speed
$N(R_N)$	Cavitation nuclei number density distribution function
$NPSP$	Net positive suction pressure
$NPSE$	Net positive suction energy
$NPSH$	Net positive suction head
p	Pressure
p_A	Radiated acoustic pressure
p^T	Total pressure
p_G	Partial pressure of gas
p_S	Sound pressure level
p_V	Vapor pressure

P	Power
$\tilde{q}^n$	Vector of fluctuating quantities
Q	Volume flow rate (or heat)
$\mathcal{Q}$	Rate of heat addition
r	Radial coordinate in turbomachine
R	Radial dimension in turbomachine
R	Bubble radius
R	Resistance
R_N	Cavitation nucleus radius
Re	Reynolds number
s	Coordinate measured in the direction of flow
s	Solidity
$\mathcal{S}$	Surface tension of the saturated vapor/liquid interface
S	Suction specific speed
S_i	Inception suction specific speed
S_a	Fractional head loss suction specific speed
S_b	Breakdown suction specific speed
Sf	Slip factor
t	Time
T	Temperature or torque
T_{ij}	Transfer matrix elements
$[T]$	Transfer matrix based on $\tilde{p}^T, \tilde{m}$
$[T^*]$	Transfer matrix based on $\tilde{p}, \tilde{m}$
$[TP]$	Pump transfer matrix
$[TS]$	System transfer matrix
u	Velocity in the s or x directions
u_i	Velocity vector
U	Fluid velocity
U_∞	Velocity of upstream uniform flow
v	Fluid velocity in non-rotating frame
V	Volume or fluid velocity
w	Fluid velocity in rotating frame
$\dot{W}$	Rate of work done on the fluid
z	Elevation
Z_{CF}	Common factor of Z_R and Z_S
Z_R	Number of rotor blades
Z_S	Number of stator blades

Greek letters

α	Angle of incidence

α_L	Thermal diffusivity of liquid
β	Angle of relative velocity vector
β_b	Blade angle relative to cross-plane
γ_n	Wave propagation speed
Γ	Geometric constant
δ	Deviation angle at flow discharge
δ	Clearance
ϵ	Eccentricity
ϵ	Angle of turn
η	Efficiency
θ	Angular coordinate
θ_c	Camber angle
θ^*	Momentum thickness of a blade wake
Θ	Thermal term in the Rayleigh-Plesset equation
ϑ	Inclination of discharge flow to the axis of rotation
κ	Bulk modulus of the liquid
μ	Dynamic viscosity
ν	Kinematic viscosity
ρ	Density of fluid
σ	Cavitation number
σ_i	Cavitation inception number
σ_a	Fractional head loss cavitation number
σ_b	Breakdown cavitation number
σ_c	Choked cavitation number
σ_{TH}	Thoma cavitation factor
Σ	Thermal parameter for bubble growth
$\Sigma_{1,2,3}$	Geometric constants
τ	Blade thickness
ϕ	Flow coefficient
ψ	Head coefficient
ψ_0	Head coefficient at zero flow
ω	Radian frequency of whirl motion or other excitation
ω_P	Bubble natural frequency
Ω	Radian frequency of shaft rotation

Subscripts

On any variable, Q:

Q_o	Initial value, upstream value or reservoir value
Q_1	Value at inlet

Q_2	Value at discharge
Q_a	Component in the axial direction
Q_b	Pertaining to the blade
Q_∞	Value far from the bubble or in the upstream flow
Q_B	Value in the bubble
Q_C	Critical value
Q_D	Design value
Q_E	Equilibrium value
Q_G	Value for the gas
Q_{H1}	Value at the inlet hub
Q_{H2}	Value at the discharge hub
Q_i	Components of vector Q
Q_i	Pertaining to a section, i, of the hydraulic system
Q_L	Saturated liquid value
Q_m	Meridional component
Q_M	Mean or maximum value
Q_N	Nominal conditions or pertaining to nuclei
Q_n, Q_t	Components normal and tangential to whirl orbit
Q_P	Pertaining to the pump
Q_r	Component in the radial direction
Q_s	Component in the s direction
Q_{T1}	Value at the inlet tip
Q_{T2}	Value at the discharge tip
Q_V	Saturated vapor value
Q_x, Q_y	Components in the x and y directions
Q_θ	Component in the circumferential (or θ) direction

Superscripts and other qualifiers

On any variable, Q:

$\bar{Q}$	Mean value of Q or complex conjugate of Q
$\tilde{Q}$	Complex amplitude of Q
$\dot{Q}$	Time derivative of Q
$\ddot{Q}$	Second time derivative of Q
Q^*	Rotordynamics: denotes dimensional Q
$Re\{Q\}$	Real part of Q
$Im\{Q\}$	Imaginary part of Q

1

Introduction

1.1 Subject

The subject of this monograph is the fluid dynamics of liquid turbomachines, particularly pumps. Rather than attempt a general treatise on turbomachines, we shall focus attention on those special problems and design issues associated with the flow of liquid through a rotating machine. There are two characteristics of a liquid that lead to these special problems, and cause a significantly different set of concerns than would occur in, say, a gas turbine. These are the potential for cavitation and the high density of liquids that enhances the possibility of damaging unsteady flows and forces.

1.2 Cavitation

The word cavitation refers to the formation of vapor bubbles in regions of low pressure within the flow field of a liquid. In some respects, cavitation is similar to boiling, except that the latter is generally considered to occur as a result of an increase of temperature rather than a decrease of pressure. This difference in the direction of the state change in the phase diagram is more significant than might, at first sight, be imagined. It is virtually impossible to cause any rapid uniform change in temperature throughout a finite volume of liquid. Rather, temperature change most often occurs by heat transfer through a solid boundary. Hence, the details of the boiling process generally embrace the detailed interaction of vapor bubbles with a solid surface, and the thermal boundary layer on that surface. On the other hand, a rapid, uniform change in pressure in a liquid is commonplace and, therefore, the details of the cavitation process may differ considerably from those that occur in boiling. Much more detail on the process of cavitation is included in later sections.

It is sufficient at this juncture to observe that cavitation is generally a malevolent process, and that the deleterious consequences can be divided into three categories. First, cavitation can cause damage to the material surfaces close to the area where the bubbles collapse when they are convected into regions of higher pressure. Cavitation damage can be very expensive, and very difficult to eliminate. For most designers of hydraulic machinery, it is the preeminent problem associated with cavitation. Frequently, one begins with the objective of eliminating cavitation completely. However, there are many circumstances in which this proves to be impossible, and the effort must be redirected into minimizing the adverse consequences of the phenomenon.

The second adverse effect of cavitation is that the performance of the pump, or other hydraulic device, may be significantly degraded. In the case of pumps, there is generally a level of inlet pressure at which the performance will decline dramatically, a phenomenon termed cavitation breakdown. This adverse effect has naturally given rise to changes in the design of a pump so as to minimize the degradation of the performance; or, to put it another way, to optimize the performance in the presence of cavitation. One such design modification is the addition of a cavitating inducer upstream of the inlet to a centrifugal or mixed flow pump impeller. Another example is manifest in the blade profiles used for supercavitating propellers. These supercavitating hydrofoil sections have a sharp leading edge, and are shaped like curved wedges with a thick, blunt trailing edge.

The third adverse effect of cavitation is less well known, and is a consequence of the fact that cavitation affects not only the steady state fluid flow, but also the unsteady or dynamic response of the flow. This change in the dynamic performance leads to instabilities in the flow that do not occur in the absence of cavitation. Examples of these instabilities are "rotating cavitation," which is somewhat similar to the phenomenon of rotating stall in a compressor, and "auto-oscillation," which is somewhat similar to compressor surge. These instabilities can give rise to oscillating flow rates and pressures that can threaten the structural integrity of the pump or its inlet or discharge ducts. While a complete classification of the various types of unsteady flow arising from cavitation has yet to be constructed, we can, nevertheless, identify a number of specific types of instability, and these are reviewed in later chapters of this monograph.

1.3 Unsteady Flows

While it is true that cavitation introduces a special set of fluid-structure interaction issues, it is also true that there are many such unsteady flow problems which can arise even in the absence of cavitation. One reason these issues may be more critical in a liquid turbomachine is that the large density of a liquid implies much larger fluid dynamic forces. Typically, fluid dynamic forces scale like $\rho \Omega^2 D^4$ where ρ is the fluid density, and Ω and D are the typical frequency of rotation and the typical length, such as the span or chord of the impeller blades or the diameter of the impeller. These forces are applied to blades whose typical thickness is denoted by τ. It follows that the typical structural stresses in the blades are given by $\rho \Omega^2 D^4/\tau^2$, and, to minimize structural problems, this quantity will have an upper bound which will depend on the material. Clearly this limit will be more stringent when the density of the fluid is larger. In many pumps and liquid turbines it requires thicker blades (larger τ) than would be advisable from a purely hydrodynamic point of view.

This monograph presents a number of different unsteady flow problems that are of concern in the design of hydraulic pumps and turbines. For example, when a rotor blade passes through the wake of a stator blade (or vice versa), it will encounter an unsteady load which is endemic to all turbomachines. Recent investigations of these loads will be reviewed. This rotor-stator interaction problem is an example of a local unsteady flow phenomenon. There also exist global unsteady flow problems, such as the auto-oscillation problem mentioned earlier. Other global unsteady flow problems are caused by the fluid-induced radial loads on an impeller due to flow asymmetries, or the fluid-induced rotordynamic loads that may increase or decrease the critical whirling speeds of the shaft system. These last issues have only recently been addressed from a fundamental research perspective, and a summary of the conclusions is included in this monograph.

1.4 Trends in Hydraulic Turbomachinery

Though the constraints on a turbomachine design are as varied as the almost innumerable applications, there are a number of ubiquitous trends which allow us to draw some fairly general conclusions. To do so we make use of the affinity laws that are a consequence of dimensional analysis, and relate performance characteristics to the density of the fluid, ρ, the typical rotational speed, Ω, and the typical diameter, D, of the pump. Thus the volume flow rate through the pump, Q, the total head rise across the pump, H, the torque, T, and the power absorbed by the pump, P, will scale according to

$$Q \propto \Omega D^3 \tag{1.1}$$

$$H \propto \Omega^2 D^2 \tag{1.2}$$

$$T \propto \rho D^5 \Omega^2 \tag{1.3}$$

$$P \propto \rho D^5 \Omega^3 \tag{1.4}$$

These simple relations allow basic scaling predictions and initial design estimates. Furthermore, they permit consideration of optimal characteristics, such as the power density which, according to the above, should scale like $\rho D^2 \Omega^3$.

One typical consideration arising out of the affinity laws relates to optimizing the design of a pump for a particular power level, P, and a particular fluid, ρ. This fixes the value of $D^5 \Omega^3$. If one wished to make the pump as small as possible (small D) to reduce weight (as is critical in the rocket engine context) or to reduce cost, this would dictate not only a higher rotational speed, Ω, but also a higher impeller tip speed, $\Omega D/2$. However, as we shall see in the next chapter, the propensity for cavitation increases as a parameter called the cavitation number decreases, and the cavitation number is inversely proportional to the square of the tip speed or $\Omega^2 D^2/4$. Consequently, the increase in tip speed suggested above could lead to a cavitation problem. Often, therefore, one designs the smallest pump that will still operate without cavitation, and this implies a particular size and speed for the device.

Furthermore, as previously mentioned, the typical fluid-induced stresses in the structure will be given by $\rho \Omega^2 D^4 / \tau^2$, and, if $D^5 \Omega^3$ is fixed and if one maintains the same geometry, D/τ, then the stresses will increase like $D^{-4/3}$ as the size, D, is decreased. Consequently, fluid/structure interaction problems will increase. To counteract this the blades are often made thicker (D/τ is decreased), but this usually leads to a decrease in the hydraulic performance of the turbomachine. Consequently an optimal design often requires a balanced compromise between hydraulic and structural requirements. Rarely does one encounter a design in which this compromise is optimal.

Of course, the design of a pump, compressor or turbine involves many factors other than the technical issues discussed above. Many compromises and engineering judgments must be made based on constraints such as cost, reliability and the expected life of a machine. This book will not attempt to deal with such complex issues, but will simply focus on the advances in the technical data base associated with cavitation and unsteady flows. For a broader perspective on the design issues, the reader is referred to engineering texts such as those listed at the end of this chapter.

1.5 Book Structure

The intention of this monograph is to present an account of both the cavitation issues and the unsteady flow issues, in the hope that this will help in the design of more effective liquid turbomachines. In chapter 2 we review some of the basic principles of the fluid mechanical design of turbomachines for incompressible fluids, and follow that, in chapter 3, with a discussion of the two-dimensional performance analyses based on the flows through cascades of foils. A brief review of three-dimensional effects and secondary flows follows in chapter 4. Then, in chapter 5, we introduce the parameters which govern the phenomenon of cavitation, and describe the different forms which cavitation can take. This is followed by a discussion of the factors which influence the onset or inception of cavitation. Chapter 6 introduces concepts from the analyses of bubble dynamics, and relates those ideas to two of the byproducts of the phenomenon, cavitation damage and noise. The isssues associated with the performance of a pump under cavitating conditions are addressed in chapter 7.

The last three chapters deal with unsteady flows and vibration in pumps. Chapter 8 presents a survey of some of the vibration problems in pumps. Chapter 9 provides details of the two basic approaches to the analysis of instabilites and unsteady flow problems in hydraulic systems, namely the methods of solution in the time domain and in the frequency domain. Where possible, it includes a survey of the existing information on the dynamic response of pumps under cavitating and non-cavitating conditions. The final chapter 10 deals with the particular fluid/structure interactions associated with rotordynamic shaft vibrations, and elucidates the fluid-induced rotordynamic forces that can result from the flows through seals and through and around impellers.

2

Basic Principles

2.1 Geometric Notation

The geometry of a generalized turbomachine rotor is sketched in figure 2.1, and consists of a set of rotor blades (number = Z_R) attached to a hub and operating within a static casing. The radii of the inlet blade tip, inlet blade hub, discharge blade tip, and discharge blade hub are denoted by R_{T1}, R_{H1}, R_{T2}, and R_{H2}, respectively. The discharge blade passage is inclined to the axis of rotation at an angle, ϑ, which would be close to 90° in the case of a centrifugal pump, and much smaller in the case of an axial flow machine. In practice, many pumps and turbines are of the "mixed flow" type, in which the typical or mean discharge flow is at some intermediate angle, $0 < \vartheta < 90°$.

The flow through a general rotor is normally visualized by developing a meridional surface (figure 2.2), that can either correspond to an axisymmetric streamsurface, or be some estimate thereof. On this meridional surface (see figure 2.2) the fluid velocity in a non-rotating coordinate system is denoted by $v(r)$ (with subscripts 1 and 2 denoting particular values at inlet and discharge) and the corresponding velocity relative to the rotating blades is denoted by $w(r)$. The velocities, v and w, have components v_θ and w_θ in the circumferential direction, and v_m and w_m in the meridional direction. Axial and radial components are denoted by the subscripts a and r. The velocity of the blades is Ωr. As shown in figure 2.2, the flow angle $\beta(r)$ is defined as the angle between the relative velocity vector in the meridional plane and a plane perpendicular to the axis of rotation. The blade angle $\beta_b(r)$ is defined as the inclination of the tangent to the blade in the meridional plane and the plane perpendicular to the axis of rotation. If the flow is precisely parallel to the blades, $\beta = \beta_b$. Specific values of the blade angle at the leading and trailing edges (1 and 2) and at the hub and tip (H and T) are denoted by the corresponding suffices, so that, for example, β_{bT2} is the blade angle at the discharge tip.

At the leading edge it is important to know the angle $\alpha(r)$ with which the flow meets the blades, and, as defined in figure 2.3,

$$\alpha(r) = \beta_{b1}(r) - \beta_1(r). \tag{2.1}$$

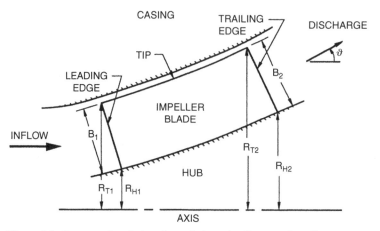

Figure 2.1. Cross-sectional view through the axis of a pump impeller.

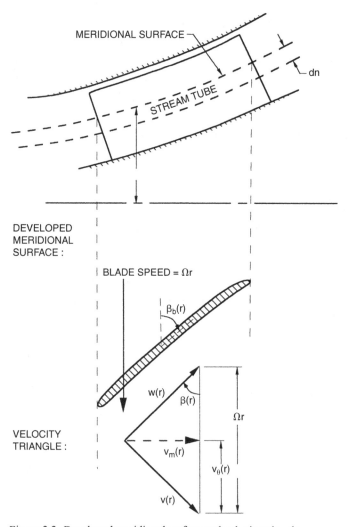

Figure 2.2. Developed meridional surface and velocity triangle.

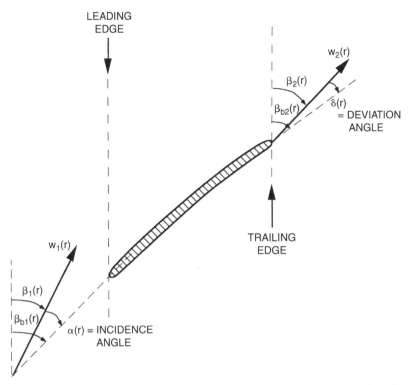

Figure 2.3. Repeat of figure 2.2 showing the definitions of the incidence angle at the leading edge and the deviation angle at the trailing edge.

This angle, α, is called the incidence angle, and, for simplicity, we shall denote the values of the incidence angle at the tip, $\alpha(R_{T1})$, and at the hub, $\alpha(R_{H1})$, by α_T and α_H, respectively. Since the inlet flow can often be assumed to be purely axial ($v_1(r) = v_{a1}$ and parallel with the axis of rotation), it follows that $\beta_1(r) = \tan^{-1}(v_{a1}/\Omega r)$, and this can be used in conjunction with equation 2.1 in evaluating the incidence angle for a given flow rate.

The incidence angle should not be confused with the "angle of attack," which is the angle between the incoming relative flow direction and the chord line (the line joining the leading edge to the trailing edge). Note, however, that, in an axial flow pump with straight helicoidal blades, the angle of attack is equal to the incidence angle.

At the trailing edge, the difference between the flow angle and the blade angle is again important. To a first approximation one often assumes that the flow is parallel to the blades, so that $\beta_2(r) = \beta_{b2}(r)$. A departure from this idealistic assumption is denoted by the deviation angle, $\delta(r)$, where, as shown in figure 2.3:

$$\delta(r) = \beta_{b2}(r) - \beta_2(r) \tag{2.2}$$

This is normally a function of the ratio of the width of the passage between the blades to the length of the same passage, a geometric parameter known as the solidity which is

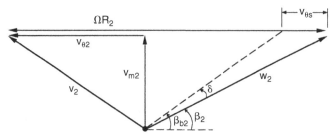

Figure 2.4. Velocity vectors at discharge indicating the slip velocity, $v_{\theta s}$.

defined more precisely below. Other angles, that are often used, are the angle through which the flow is turned, known as the *deflection angle*, $\beta_2 - \beta_1$, and the corresponding angle through which the blades have turned, known as the *camber angle* and denoted by $\theta_c = \beta_{b2} - \beta_{b1}$.

Deviation angles in radial machines are traditionally represented by the *slip velocity*, $v_{\theta s}$, which is the difference between the actual and ideal circumferential velocities of the discharge flow, as shown in figure 2.4. It follows that

$$v_{\theta s} = \Omega R_2 - v_{\theta 2} - v_{r2} \cot \beta_{b2} \qquad (2.3)$$

This, in turn, is used to define a parameter known as the *slip factor*, Sf, where

$$Sf = 1 - \frac{v_{\theta s}}{\Omega R_2} = 1 - \phi_2 (\cot \beta_2 - \cot \beta_{b2}) \qquad (2.4)$$

Other, slightly different "slip factors" have also been used in the literature; for example, Stodola (1927), who originated the concept, defined the slip factor as $1 - v_{\theta s}/\Omega R_2(1 - \phi_2 \cot \beta_{b2})$. However, the definition 2.4 is now widely used. It follows that the deviation angle, δ, and the slip factor, Sf, are related by

$$\delta = \beta_{b2} - \cot^{-1}\left(\cot \beta_{b2} + \frac{(1 - Sf)}{\phi_2}\right) \qquad (2.5)$$

where the flow coefficient, ϕ_2, is defined later in equation 2.17.

2.2 Cascades

We now turn to some specific geometric features that occur frequently in discussions of pumps and other turbomachines. In a purely axial flow machine, the development of a cylindrical surface within the machine produces a *linear cascade* of the type shown in figure 2.5(a). The centerplane of the blades can be created using a "generator," say $z = z^*(r)$, which is a line in the rz-plane. If this line is rotated through a helical path, it describes a helicoidal surface of the form

$$z = z^*(r) + \frac{h_p \theta}{2\pi} \qquad (2.6)$$

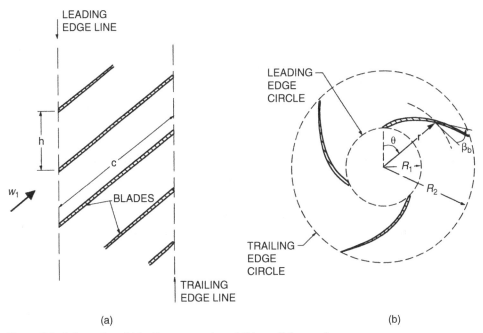

Figure 2.5. Schematics of (a) a linear cascade and (b) a radial cascade.

where h_p is the "pitch" of the helix. Of course, in many machines, the pitch is also a function of θ so that the flow is turned by the blades. If, however, the pitch is constant, the development of a cylindrical surface will yield a cascade with straight blades and constant blade angle, β_b. Moreover, the blade thickness is often neglected, and the blades in figure 2.5(a) then become infinitely thin lines. Such a cascade of infinitely thin, flat blades is referred to as a *flat plate cascade*.

It is convenient to use the term "simple" cascade to refer to those geometries for which the blade angle, β_b, is constant whether in an axial, radial, or mixed flow machine. Clearly, the flat plate cascade is the axial flow version of a simple cascade.

Now compare the geometries of the cascades at different radii within an axial flow machine. Later, we analyse the cavitating flow occurring at different radii (see figure 7.35). Often the pitch at a given axial position is the same at all radii. Then it follows that the radial variation in the blade angle, $\beta_b(r)$, must be given by

$$\beta_b(r) = \tan^{-1}\left[\frac{R_T \tan \beta_{bT}}{r}\right] \tag{2.7}$$

where β_{bT} is the blade angle at the tip, $r = R_T$.

In a centrifugal machine in which the flow is purely radial, a cross-section of the flow would be as shown in figure 2.5(b), an array known as a *radial cascade*. In a *simple* radial cascade, the angle, β_b, is uniform along the length of the blades. The resulting blade geometry is known as a logarithmic spiral, since it follows that the coordinates of the blades are given by the equation

$$\theta - \theta_0 = A \ln r \tag{2.8}$$

where $A = \cot \beta_b$ and θ_0 are constants. Logarithmic spiral blades are therefore equivalent to straight blades in a linear cascade. Note that a fluid particle in a flow of uniform circulation and constant source strength at the origin will follow a logarithmic spiral since all velocities will be of the form C/r where C is a uniform constant.

In any of type of pump, the ratio of the length of a blade passage to its width is important in determining the degree to which the flow is guided by the blades. The solidity, s, is the geometric parameter that is used as a measure of this geometric characteristic, and s can be defined for any *simple* cascade as follows. If we identify the difference between the θ coordinates for the same point on adjacent blades (call this $\Delta\theta_A$) and the difference between the θ coordinates for the leading and trailing edges of a blade (call this $\Delta\theta_B$), then the solidity for a simple cascade is defined by

$$s = \frac{\Delta\theta_B}{\Delta\theta_A \cos \beta_b} \tag{2.9}$$

Applying this to the linear cascade of figure 2.5(a), we find the familiar

$$s = c/h \tag{2.10}$$

In an axial flow pump this corresponds to $s = Z_R c/2\pi R_{T1}$, where c is the chord of the blade measured in the developed meridional plane of the blade tips. On the other hand, for the radial cascade of figure 2.5(b), equation 2.9 yields the following expression for the solidity:

$$s = Z_R \ell n \left(R_2/R_1 \right) / 2\pi \sin \beta_b \tag{2.11}$$

which is, therefore, geometrically equivalent to c/h in the linear cascade.

In practice, there exist many "mixed flow" pumps whose geometries lie between that of an axial flow machine ($\vartheta = 0$, figure 2.1) and that of a radial machine ($\vartheta = \pi/2$). The most general analysis of such a pump would require a cascade geometry in which figures 2.5(a) and 2.5(b) were *projections* of the geometry of a meridional surface (figure 2.2) onto a cylindrical surface and onto a plane perpendicular to the axis, respectively. (Note that the β_b marked in figure 2.5(b) is not appropriate when that diagram is used as a projection). We shall not attempt such generality here; rather, we observe that the meridional surface in many machines is close to conical. Denoting the inclination of the cone to the axis by ϑ, we can use equation 2.9 to obtain an expression for the solidity of a simple cascade in this conical geometry,

$$s = Z_R \ell n \left(R_2/R_1 \right) / 2\pi \sin \beta_b \sin \vartheta \tag{2.12}$$

Clearly, this includes the expressions 2.10 and 2.11 as special cases.

2.3 Flow Notation

The flow variables that are important are, of course, the static pressure, p, the total pressure, p^T, and the volume flow rate, Q. Often the total pressure is defined by the total head, $p^T/\rho g$. Moreover, in most situations of interest in the context of turbomachinery, the potential energy associated with the earth's gravitational field is negligible relative to the kinetic energy of the flow, so that, by definition

$$p^T = p + \frac{1}{2}\rho v^2 \tag{2.13}$$

$$p^T = p + \frac{1}{2}\rho \left(v_m^2 + v_\theta^2\right) \tag{2.14}$$

$$p^T = p + \frac{1}{2}\rho \left(w^2 + 2r\Omega v_\theta - \Omega^2 r^2\right) \tag{2.15}$$

using the velocity triangle of figure 2.2. In an incompressible flow, the total pressure represents the total mechanical energy per unit volume of fluid, and, therefore, the change in total pressure across the pump, $p_2^T - p_1^T$, is a fundamental measure of the mechanical energy imparted to the fluid by the pump.

It follows that, in a pump with an incompressible fluid, the overall characteristics that are important are the volume flow rate, Q, and the total pressure rise, $\rho g H$, where $H = (p_2^T - p_1^T)/\rho g$ is the total head rise. These dimensional characteristics are conveniently nondimensionalized by defining a head coefficient, ψ,

$$\psi = (p_2^T - p_1^T)/\rho R_{T2}^2 \Omega^2 = g H/R_{T2}^2 \Omega^2 \tag{2.16}$$

and one of two alternative flow coefficients, ϕ_1 and ϕ_2:

$$\phi_1 = Q/A_1 R_{T1}\Omega \quad \text{or} \quad \phi_2 = Q/A_2 R_{T2}\Omega \tag{2.17}$$

where A_1 and A_2 are the inlet and discharge areas, respectively. The discharge flow coefficient is the nondimensional parameter most often used to describe the flow rate. However, in discussions of cavitation, which occurs at the inlet to a pump impeller, the inlet flow coefficient is a more sensible parameter. Note that, for a purely axial inflow, the incidence angle is determined by the flow coefficient, ϕ_1:

$$\alpha(r) = \beta_{b1}(r) - \tan^{-1}(\phi_1 r/R_{T1}) \tag{2.18}$$

Furthermore, for a given deviation angle, specifying ϕ_2 fixes the geometry of the velocity triangle at discharge from the pump.

Frequently, the conditions at inlet and/or discharge are nonuniform and one must subdivide the flow into annular streamtubes, as indicated in figure 2.2. Each streamtube must then be analysed separately, using the blade geometry pertinent at that radius. The mass flow rate, m, through an individual streamtube is given by

$$m = 2\pi \rho r v_m dn \tag{2.19}$$

where n is a coordinate measured normal to the meridional surface, and, in the present text, will be useful in describing the discharge geometry.

Conservation of mass requires that m have the same value at inlet and discharge. This yields a relation between the inlet and discharge meridional velocities, that involves the cross-sectional areas of the streamtube at these two locations. The total volume flow rate through the turbomachine, Q, is then related to the velocity distribution at any location by the integral

$$Q = \int 2\pi r v_m(r) dn \tag{2.20}$$

The total head rise across the machine, H, is given by the integral of the total rate of work done on the flow divided by the total mass flow rate:

$$H = \frac{1}{Q} \int \frac{(p_2^T(r) - p_1^T(r))}{\rho g} 2\pi r v_m(r) dn \tag{2.21}$$

These integral expressions for the flow rate and head rise will be used in later chapters.

2.4 Specific Speed

At the beginning of any pump design process, neither the size nor the shape of the machine is known. The task the pump is required to perform is to use a shaft rotating at a frequency, Ω (in rad/s), to pump a certain flow rate, Q (in m^3/s) through a head rise, H (in m). As in all fluid mechanical formulations, one should first seek a nondimensional parameter (or parameters) which distinguishes the nature of this task. In this case, there is one and only one nondimensional parametric group that is appropriate and this is known as the "specific speed," denoted by N. The form of the specific speed is readily determined by dimensional analysis:

$$N = \frac{\Omega Q^{\frac{1}{2}}}{(gH)^{\frac{3}{4}}} \tag{2.22}$$

Though originally constructed to allow evaluation of the shaft speed needed to produce a particular head and flow, the name "specific speed" is slightly misleading, because N is just as much a function of flow rate and head rise as it is of shaft speed. Perhaps a more general name, like "the basic performance parameter," would be more appropriate. Note that the specific speed is a size-independent parameter, since the size of the machine is not known at the beginning of the design process.

The above definition of the specific speed has employed a consistent set of units, so that N is truly dimensionless. With these consistent units, the values of N for most common turbomachines lie in the range between 0.1 and 4.0 (see below). Unfortunately, it has been traditional in industry to use an inconsistent set of units in calculating

N. In the USA, the g is dropped from the denominator, and values for the speed, flow rate, and head in rpm, gpm, and ft are used in calculating N. This yields values that are a factor of 2734.6 larger than the values of N obtained using consistent units. The situation is even more confused since the Europeans use another set of inconsistent units (rpm, m^3/s, head in m, and no g) while the British employ a definition similar to the United States, but with Imperial gallons rather than U.S. gallons. One can only hope that the pump (and turbine) industries would cease the use of these inconsistent measures that would be regarded with derision by any engineer outside of the industry. In this monograph, we shall use the dimensionally consistent and, therefore, universal definition of N.

Note that, since Q and gH were separately nondimensionalized in the definitions 2.16 and 2.17, N can be related to the corresponding flow and head coefficients by

$$N = \left[\frac{\pi}{\cos \vartheta} \left(1 - \frac{R_{H2}^2}{R_{T2}^2} \right) \right]^{\frac{1}{2}} \frac{\phi_2^{\frac{1}{2}}}{\psi^{\frac{3}{4}}} \qquad (2.23)$$

In the case of a purely centrifugal discharge ($\vartheta = \pi/2$), the quantity within the square brackets reduces to $2\pi B_2/R_{T2}$.

Since turbomachines are designed for specific tasks, the subscripted N_D will be used to denote the design value of the specific speed for a given machine.

2.5 Pump Geometries

Since the task specifications for a pump (or turbine or compressor or other machine) can be reduced to the single parameter, N_D, it is not surprising that the overall or global geometries of pumps, that have evolved over many decades, can be seen to fit quite neatly into a single parameter family of shapes. This family is depicted in figure 2.6. These geometries reflect the fact that an axial flow machine, whether a pump, turbine, or compressor, is more efficient at high specific speeds (high flow rate, low head) while a radial machine, that uses the centrifugal effect, is more efficient at low specific speeds (low flow rate, high head). The same basic family of geometries is presented quantitatively in figure 2.7, where the anticipated head and flow coefficients are also plotted. While the existence of this parametric family of designs has emerged almost exclusively as a result of trial and error, some useful perspectives can be obtained from an approximate analysis of the effects of the pump geometry on the hydraulic performance (see section 4.3).

Normally, turbomachines are designed to have their maximum efficiency at the design specific speed, N_D. Thus, in any graph of efficiency against specific speed, each pump geometry will trace out a curve with a maximum at its optimum specific speed, as illustrated by the individual curves in figure 2.8. Furthermore, Balje (1981) has made note of another interesting feature of this family of curves in the graph of efficiency against specific speed. First, he corrects the curves for the different viscous

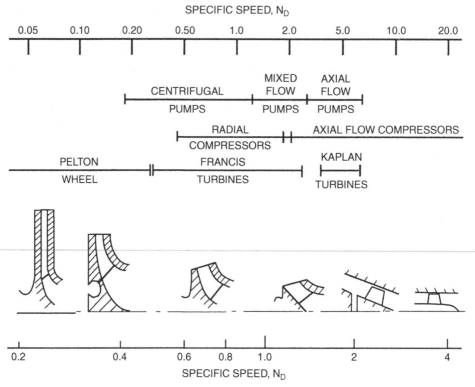

Figure 2.6. Ranges of specific speeds for typical turbomachines and typical pump geometries for different design speeds (from Sabersky, Acosta and Hauptmann 1989).

effects which can occur in machines of different size and speed, by comparing the data on efficiency at the same effective Reynolds number using the diagram reproduced as figure 2.9. Then, as can be seen in figure 2.8, the family of curves for the efficiency of different types of machines has an upper envelope with a maximum at a specific speed of unity. Maximum possible efficiencies decline for values of N_D greater or less than unity. Thus the "ideal" pump would seem to be that with a design specific speed of unity, and the maximum obtainable efficiency seems to be greatest at this specific speed. Fortunately, from a design point of view, one of the specifications has some flexibility, namely the shaft speed, Ω. Though the desired flow rate and head rise are usually fixed, it may be possible to choose the drive motor to turn at a speed, Ω, which brings the design specific speed close to the optimum value of unity.

2.6 Energy Balance

The next step in the assessment of the performance of a turbomachine is to consider the application of the first and second laws of thermodynamics to such devices. In doing so we shall characterize the inlet and discharge flows by their pressure, velocity, enthalpy, etc., assuming that these are uniform flows. It is understood that when the inlet and

2.6 Energy Balance

15

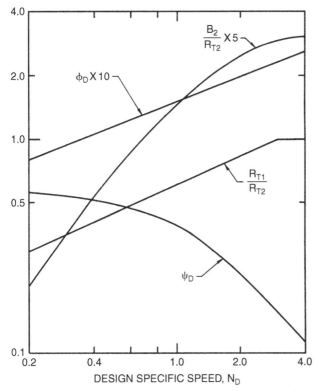

Figure 2.7. General design guidelines for pumps indicating the optimum ratio of inlet to discharge tip radius, R_{T1}/R_{T2}, and discharge width ratio, B_2/R_{T2}, for various design specific speeds, N_D. Also shown are approximate pump performance parameters, the design flow coefficient, ϕ_D, and the design head coefficient, ψ_D (adapted from Sabersky, Acosta and Hauptmann 1989).

discharge flows are non-uniform, the analysis actually applies to a single streamtube and the complete energy balance requires integration over all of the streamtubes.

The basic thermodynamic measure of the energy stored in a unit mass of flowing fluid is the total specific enthalpy (total enthalpy per unit mass) denoted by h^T and defined by

$$h^T = h + \frac{1}{2}|u|^2 + gz = e + \frac{p}{\rho} + \frac{1}{2}|u|^2 + gz \qquad (2.24)$$

where e is the specific internal energy, $|u|$ is the magnitude of the fluid velocity, and z is the vertical elevation. This expression omits any energy associated with additional external forces (for example, those due to a magnetic field), and assumes that the process is chemically inert.

Consider the steady state operation of a fluid machine in which the entering fluid has a total specific enthalpy of h_1^T, the discharging fluid has a total specific enthalpy of h_2^T, the mass flow rate is m, the net rate of heat addition to the machine is Q, and the net rate of work done on the fluid in the machine by external means is $\dot{W}$. It follows

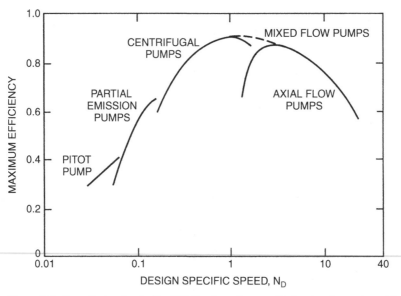

Figure 2.8. Compilation by Balje (1981) of maximum efficiencies for various kinds of pumps as a function of design specific speed, N_D. Since efficiency is also a function of Reynolds number the data has been corrected to a Reynolds number, $2\Omega R_{T2}^2/\nu$, of 10^8.

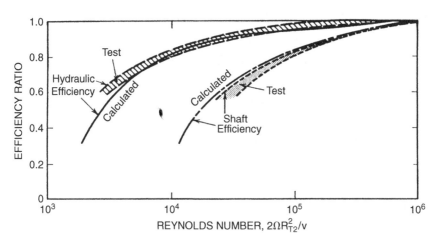

Figure 2.9. The dependence of hydraulic efficiency, η_P, and shaft efficiency, η_S, on Reynolds number, $2\Omega R_{T2}^2/\nu$ (from Balje 1981).

from the first law of thermodynamics that

$$m(h_2^T - h_1^T) = Q + \dot{W} \tag{2.25}$$

Now consider incompressible, inviscid flow. It is a fundamental property of such a flow that it contains no mechanism for an exchange of thermal and mechanical energy, and, therefore, equation 2.25 divides into two parts, governing the mechanical and thermal components of the total enthalpy, as follows:

$$(p/\rho + \frac{1}{2}|u|^2 + gz)_2 - (p/\rho + \frac{1}{2}|u|^2 + gz)_1 = \frac{(p_2^T - p_1^T)}{\rho} = \frac{\dot{W}}{m} \qquad (2.26)$$

$$e_2 - e_1 = Q/m \qquad (2.27)$$

Thus, for incompressible inviscid flow, the fluid mechanical problem (for which equation 2.26 represents the basic energy balance) can be decoupled from the heat transfer problem (for which the heat balance is represented by equation 2.27).

It follows that, if T is the torque applied by the impeller to the fluid, then the rate of work done on the fluid is $\dot{W} = T\Omega$. Consequently, in the case of an ideal fluid which is incompressible and inviscid, equation 2.26 yields a relation connecting the total pressure rise across the pump, $p_2^T - p_1^T$, the mass flow rate, m, and the torque:

$$m\frac{p_2^T - p_1^T}{\rho} = T\Omega \qquad (2.28)$$

Furthermore, the second law of thermodynamics implies that, in the presence of irreversible effects such as those caused by viscosity, the equality in equation 2.28 should be replaced by an inequality, namely a "less than" sign. Consequently, in a real pump operating with an incompressible fluid, viscous effects will cause some of the input energy to be converted to heat rather than to an increase in the stored energy in the fluid. It follows that the right-hand side of equation 2.28 is the actual work done on the fluid by the impeller, and the left-hand side is the fraction of that work which ends up as mechanical energy stored in the fluid. It is, therefore, appropriate to define a quantity, η_P, known as the pump hydraulic efficiency, to represent that fraction of the work done on the fluid that ends up as an increase in the mechanical energy stored in the fluid:

$$\eta_P = m\left(p_2^T - p_1^T\right)/\rho T\Omega \qquad (2.29)$$

Of course, additional mechanical losses may occur in a pump. These can cause the rate of work transmitted through the external shaft of the pump to be greater than the rate at which the impeller does work on the fluid. For example, losses may occur in the bearings or as a result of the "disk friction" losses caused by the fluid dynamic drag on other, non-active surfaces rotating with the shaft. Consequently, the overall (or shaft) efficiency, η_S, may be significantly smaller than η_P. For approximate evaluations of these additional losses, the reader is referred to the work of Balje (1981).

Despite all these loss mechanisms, pumps can be surprisingly efficient. A well designed centrifugal pump should have an overall efficiency in the neighborhood of 85% and some very large pumps (for example, those in the Grand Coulee Dam) can exceed 90%. Even centrifugal pumps with quite simple and crude geometries can often be 60% efficient.

2.7 Noncavitating Pump Performance

It is useful at this point to develop an approximate and idealized evaluation of the hydraulic performance of a pump in the absence of cavitation. This will take the form of an analytical expression for the head rise (or ψ) as a function of the flow rate (or ϕ_2).

To simplify this analysis it is assumed that the flow is incompressible, axisymmetric and steady in the rotating framework of the impeller blades; that the blades are infinitely thin; and that viscous losses can be neglected. Under these conditions the flow in any streamtube, such as depicted in figure 2.2, will follow the Bernoulli equation for a rotating system (see, for example, Sabersky, Acosta and Hauptmann 1989),

$$\frac{2p_1}{\rho} + w_1^2 - r_1^2\Omega^2 = \frac{2p_2}{\rho} + w_2^2 - r_2^2\Omega^2 \tag{2.30}$$

This equation can be usefully interpreted as an energy equation as follows. The terms $p + \frac{1}{2}\rho w^2$ on either side are the total pressure or mechanical energy per unit volume of fluid, and this quantity would be the same at inlet and discharge were it not for the fact that "potential" energy is stored in the rotating fluid. The term $\rho(r_1^2 - r_2^2)\Omega^2/2$ represents the difference in this "potential" energy at inlet and discharge. Clearly, when there are losses, equation 2.30 will no longer be true.

Using the definition of the total pressure (equation 2.13) and the relations between the velocities derived from the velocity triangles of figure 2.2, equation 2.30 can be manipulated to yield the following expression for the total pressure rise, $(p_2^T - p_1^T)$, for a given streamtube:

$$p_2^T - p_1^T = p_2 - p_1 + \frac{\rho}{2}\left(v_2^2 - v_1^2\right) \tag{2.31}$$

$$= \rho(\Omega r_2 v_{\theta2} - \Omega r_1 v_{\theta1}) \tag{2.32}$$

In the absence of inlet swirl ($v_{\theta1} = 0$), this leads to the nondimensional performance characteristic

$$\psi = 1 - \phi_2 \cot \beta_{b2} \tag{2.33}$$

using the definitions in equations 2.16 and 2.17. Here we have assumed that the inlet and discharge conditions are uniform which, in effect, restricts the result to a turbomachine in which the widths, B_1 and B_2 (figure 2.1), are such that $B_1 \ll R_{T1}, B_2 \ll R_{T2}$, and in which the velocities of the flow and the impeller are uniform across both the inlet and the discharge. Usually this is not the case, and the results given by equations 2.32 and 2.33 then become applicable to each individual streamtube. Integration over all the streamtubes is necessary to obtain the performance characteristic for the machine. An example of this integration was given in section 2.3. Even in these nonuniform cases, the simple expression 2.33 is widely used in combination with some mean or effective discharge blade angle, β_{b2}, to estimate the performance of a pump.

It is important to note that the above results can be connected with those of the preceding section by applying the angular momentum theorem (Newton's second law of motion applied to rotational motion) to relate the torque, T, to the net flux of angular momentum out of the pump:

$$T = m(r_2 v_{\theta 2} - r_1 v_{\theta 1}) \tag{2.34}$$

where, as before, m is the mass flow rate. Note that this momentum equation 2.34 holds whether or not there are viscous losses. In the absence of viscous losses, a second expression for the torque, T, follows from equation 2.28. By equating the two expressions, the result 2.32 for the performance in the absence of viscous losses is obtained by an alternative method.

2.8 Several Specific Impellers and Pumps

Throughout this monograph, we shall make reference to experimental data on various phenomena obtained with several specific impellers and pumps. It is appropriate at this point to include a brief description of these components. The descriptions will also serve as convenient examples of pump geometries.

Impeller X, which is shown in figure 2.10, is a 5-bladed centrifugal pump impeller made by Byron Jackson Pump Division of Borg Warner International Products. It has a discharge radius, $R_{T2} = 8.1\ cm$, a discharge blade angle, β_{bT2}, of 23°, and a design

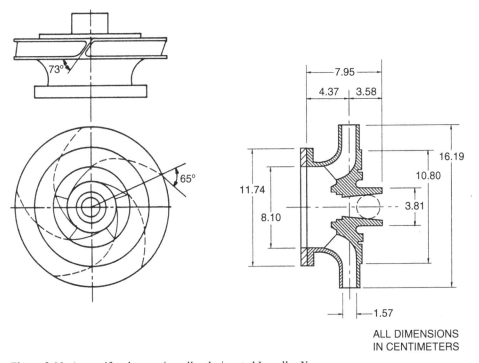

Figure 2.10. A centrifugal pump impeller designated Impeller X.

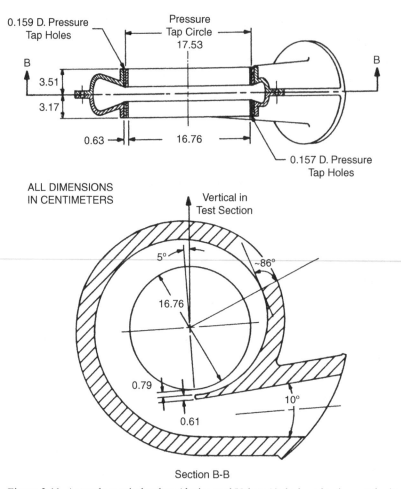

0.159 D. Pressure
Tap Holes

Pressure
Tap Circle
17.53

B   B

3.51

3.17

0.63   16.76

0.157 D. Pressure
Tap Holes

ALL DIMENSIONS
IN CENTIMETERS

Vertical in
Test Section

5°   ~86°

16.76

0.79   10°

0.61

Section B-B

Figure 2.11. A vaneless spiral volute (designated Volute A) designed to be matched to Impeller X.

specific speed, N_D, of 0.57. Impeller X was often tested in combination with Volute A (figure 2.11), a single exit, spiral volute with a base circle of 18.3 *cm* and a spiral angle of 4°. It is designed to match Impeller X at a flow coefficient of $\phi_2 = 0.092$. This implies that the principles of fluid continuity and momentum have been utilized in the design, so that the volute collects a circumferentially uniform discharge from the impeller and channels it to the discharge line in such a way that the pressure in the volute is circumferentially uniform, and in a way that minimizes the viscous losses in the decelerating flow. For given volute and impeller geometries, these objectives can only by met at one "design" flow coefficient, as described in section 4.4. We would therefore expect that the hydraulic losses would increase, and the efficiency decrease, at off-design conditions. It is valuable to emphasize that the performance of a pump depends not only on the separate designs of the impeller and volute but also on the matching of the two components.

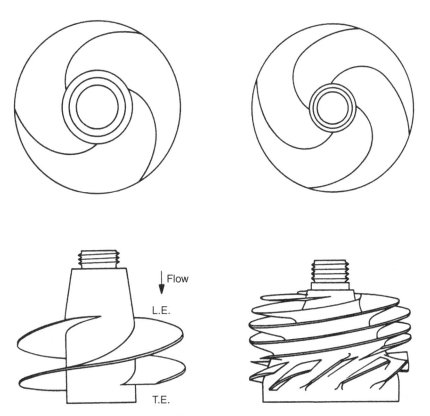

Figure 2.12. Two cavitating inducers for which performance data is presented. On the left a 7.58 *cm* diameter, 9° helical Impeller V (a 10.2 *cm* version is designated Impeller VII). On the right a 7.58 *cm* diameter scale model of the impeller in the SSME low pressure LOX turbopump, Impeller IV (a 10.2 *cm* version is designated Impeller VI).

Two particular axial flow pumps or inducers, designed to function with cavitation, will also be referred to frequently. These are shown in figure 2.12. In a number of contexts, data for several simple 9° helical inducers ($\beta_{bT1} = 9°$) will be used for illustrative purposes, and a typical geometry is shown on the left of figure 2.12. Two 7.58 *cm* diameter versions were deployed: Impeller III had straight, radial leading edges and Impeller V, with swept leading edges, is shown in figure 2.12. A 10.2 *cm* diameter version with swept leading edges is designated Impeller VII.

The second inducer geometry is pertinent to a somewhat lower specific speed. Impellers IV (7.58 *cm* diameter) and VI (10.2 *cm* diameter) were scale models of the low pressure liquid oxygen impeller in the Space Shuttle Main Engine (SSME). These have a design flow coefficient of about 0.076; other dimensions are given in table 7.1. Furthermore, some detailed data on blade angles, $\beta_{b1}(r)$, and blade thickness are given in figure 7.39.

3

Two-Dimensional Performance Analysis

3.1 Introduction

In this and the following chapter, we briefly survey the more detailed analyses of the flow in axial and centrifugal pumps, and provide a survey of some of the models used to synthesize the noncavitating performance of these turbomachines. The survey begins in this chapter with a summary of some of the results that emerge from a more detailed analysis of the two-dimensional flow in the meridional plane of the turbomachine, while neglecting most of the three-dimensional effects. In this regard, sections 3.2 through 3.4 address the analyses of linear cascades for axial flow machines, and section 3.5 summarizes the analyses of radial cascades for centrifugal machines. Three-dimensional effects are addressed in the next chapter.

3.2 Linear Cascade Analyses

The fluid mechanics of a linear cascade will now be examined in more detail, so that the role played by the geometry of the blades and information on the resulting forces on individual blades may be used to supplement the analysis of section 2.7. Referring to the periodic control volume indicated in figure 3.1, and applying the momentum theorem to this control volume, the forces, F_x and F_y, imposed by the fluid on each blade (per unit depth normal to the sketch), are given by

$$F_x = -(p_2 - p_1)h \tag{3.1}$$

$$F_y = \rho h v_m (w_1 \cos \beta_1 - w_2 \cos \beta_2) \tag{3.2}$$

where, as a result of continuity, $v_{m1} = v_{m2} = v_m$. Note that F_y is entirely consistent with the expression 2.34 for the torque, T.

To proceed, we define the vector mean of the relative velocities, w_1 and w_2, as having a magnitude w_M and a direction β_M, where by simple geometry

$$\cot \beta_M = \frac{1}{2} (\cot \beta_1 + \cot \beta_2) \tag{3.3}$$

$$w_M = v_m / \sin \beta_M \tag{3.4}$$

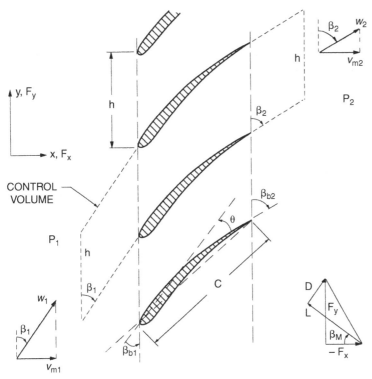

Figure 3.1. Schematic of a linear cascade showing the blade geometry, the periodic control volume and the definition of the lift, L, and drag, D, forces on a blade.

It is conventional and appropriate (as discussed below) to define the lift, L, and the drag, D, components of the total force on a blade, $(F_x^2 + F_y^2)^{\frac{1}{2}}$, as the components normal and tangential to the vector mean velocity, w_M. More specifically, as shown in figure 3.1,

$$L = -F_x \cos \beta_M + F_y \sin \beta_M \qquad (3.5)$$

$$D = F_x \sin \beta_M + F_y \cos \beta_M \qquad (3.6)$$

where L and D are forces per unit depth normal to the sketch. Nondimensional lift and drag coefficients are defined as

$$C_L = L \Big/ \frac{1}{2}\rho w_M^2 c; \quad C_D = D \Big/ \frac{1}{2}\rho w_M^2 c \qquad (3.7)$$

The list of fundamental relations is complete if we write the expression for the pressure difference across the cascade as

$$p_1 - p_2 = \Delta p_L^T + \frac{\rho}{2}\left(w_1^2 - w_2^2\right) \qquad (3.8)$$

where Δp_L^T denotes the total pressure loss across the cascade caused by viscous effects. In frictionless flow, $\Delta p_L^T = 0$, and the relation 3.8 becomes the Bernoulli equation in rotating coordinates (equation 2.30 with $r_1 = r_2$ as is appropriate here). A nondimensional loss coefficient, f, is defined as:

$$f = \Delta p_L^T / \frac{1}{2} \rho w_M^2 \tag{3.9}$$

Equations 3.1 through 3.9 can be manipulated to obtain expressions for the lift and drag coefficients as follows:

$$C_D = 2f \sin \beta_M / s \tag{3.10}$$

$$C_L = \frac{2}{s} \left[\frac{\psi}{\phi} \sin \beta_M + \frac{f(\phi - \cos \beta_M \sin \beta_M)}{\sin \beta_M} \right] \tag{3.11}$$

where $s = c/h$ is the solidity, ψ is the head coefficient, $(p_2^T - p_1^T)/\rho \Omega^2 R^2$, and ϕ is the flow coefficient, $v_m/\Omega R$. Note that in frictionless flow $C_D = 0$ and $C_L = 2\psi \sin \beta_M /\phi s$; then the total force (lift) on the foil is perpendicular to the direction defined by the β_M of equation 3.3. This provides confirmation that the directions we chose in defining L and D (see figure 3.1) were appropriate for, in frictionless flow, C_D must indeed be zero.

Also note that equations 3.1 through 3.9 yield the head/flow characteristic given by

$$\psi = \phi (\cot \beta_1 - \cot \beta_2) - f\phi^2 \left(1 + \cot^2 \beta_M\right) \tag{3.12}$$

which, when there is no inlet swirl or prerotation so that $\tan \beta_1 = \phi$, becomes

$$\psi = 1 - \phi \cot \beta_2 - f \left[\phi^2 + \frac{1}{4}(1 + \phi \cot \beta_2)^2 \right] \tag{3.13}$$

In frictionless flow, when the discharge is parallel with the blades ($\beta_2 = \beta_{b2}$), this, of course, reduces to the characteristic equation 2.33. Note that the use of the relation 3.13 allows us to write the expression 3.11 for the lift coefficient as

$$C_L = \frac{2}{s} [2 \sin \beta_M (\cot \beta_1 - \cot \beta_M) - f \cos \beta_M] \tag{3.14}$$

Figure 3.2 presents examples of typical head/flow characteristics resulting from equation 3.13 for some chosen values of β_2 and the friction coefficient, f. It should be noted that, in any real turbomachine, f will not be constant but will vary substantially with the flow coefficient, ϕ, which determines the angle of incidence and other flow characteristics. More realistic cases are presented a little later in figure 3.3.

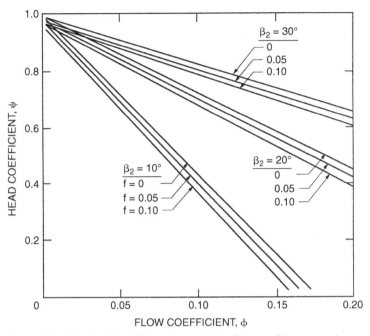

Figure 3.2. Calculated head/flow characteristics for some linear cascades.

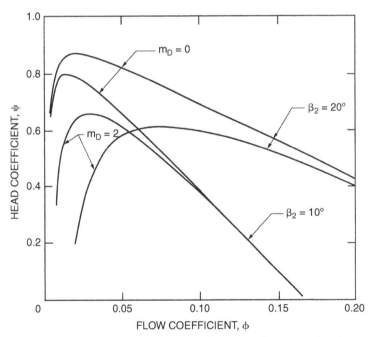

Figure 3.3. Calculated head/flow characteristics for a linear cascade using blade drag coefficients given by equation 3.18 with $C_{D0} = 0.02$. The corresponding characteristics with $C_{D0} = m_D = 0$ are shown in figure 3.2.

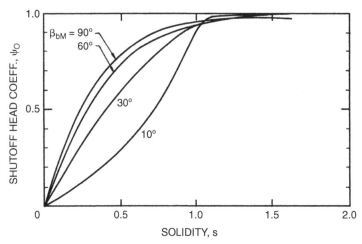

Figure 3.4. The performance parameter, ψ_0, as a function of solidity, s, for flat plate cascades with different blade angles, β_b. Adapted by Wislicensus (1947) (see also Sabersky, Acosta and Hauptmann 1989) from the potential flow theory of König (1922).

The observant reader will have noted that all of the preceding equations of this section involve only the inclinations of the flow and *not* of the blades, which have existed only as ill-defined objects that achieve the turning of the flow. In order to progress further, it is necessary to obtain a detailed solution of the flow, one result of which will be the connection between the flow angles (β_M, β_2) and the geometry of the blades, including the blade angles (β_b, β_{b1}, β_{b2}). A large literature exists describing methods for the solutions of these flows, but such detail is beyond the scope of this text. As in most high Reynolds number flows, one begins with potential flow solutions, for which the reader should consult a modern text, such as that by Horlock (1973), or the valuable review by Roudebush (1965). König (1922) produced one of the earliest potential flow solutions, namely that for a simple flat plate cascade of infinitely thin blades. This was used to generate figure 3.4. Such potential flow methods must be supplemented by viscous analyses of the boundary layers on the blades and the associated wakes in the discharge flow. Leiblein (1965) provided an excellent review of these viscous flow methods, and some of his basic methodology will be introduced later.

To begin with, however, one can obtain some useful insights by employing our basic knowledge and understanding of lift and drag coefficients obtained from tests, both those on single blades (airfoils, hydrofoils) and those on cascades of blades. One such observation is that the lift coefficient, C_L, is proportional to the sine of the angle of attack, where the angle of attack is defined as the angle between the mean flow direction, β_M, and a mean blade angle, β_{bM}. Thus

$$C_L = m_L \sin(\beta_{bM} - \beta_M) \tag{3.15}$$

where m_L is a constant, a property of the blade or cascade geometry. In the case of frictionless flow ($f = 0$), the expression 3.15 may be substituted into equation 3.14, resulting in an expression for β_M. When this is used with equation 3.13, the following head/flow characteristic results:

$$\psi = \frac{2m_L s \sin \beta_{bM}}{4 + m_L s \sin \beta_{bM}} \left[1 - \phi \left(\cot \beta_{bM} + \frac{v_{\theta 1}}{v_{m1}} \right) \right] \tag{3.16}$$

where, for convenience, the first factor on the right-hand side is denoted by

$$\psi_0 = \frac{2m_L s \sin \beta_{bM}}{4 + m_L s \sin \beta_{bM}} = \left[1 + \frac{\cot \beta_2 - \cot \beta_{b2}}{\cot \beta_1 - \cot \beta_2} \right]^{-1} \tag{3.17}$$

The factor, ψ_0, is known as the frictionless shut-off head coefficient, since it is equal to the head coefficient at zero flow rate. The second expression for ψ_0 follows from the preceding equations, and will be used later. Note that, unlike equation 3.13, the head/flow characteristic of equation 3.16 is given in terms of m_L and practical quantities, such as the blade angle, β_{bM}, and the inlet swirl or prerotation, $v_{\theta 1}/v_{m1}$.

It is also useful to consider the drag coefficient, C_D, for it clearly defines f and the viscous losses in the cascade. Instead of being linear with angle of attack, C_D will be an even function so an appropriate empirical result corresponding to equation 3.15 would be

$$C_D = C_{D0} + m_D \sin^2 (\beta_{bM} - \beta_M) \tag{3.18}$$

where C_{D0} and m_D are constants. Some head/flow characteristics resulting from typical values of C_{D0} and m_D are shown in figure 3.3. Note that these performance curves have a shape that is closer to practical performance curves than the constant friction factor results of figure 3.2.

3.3 Deviation Angle

While the simple, empirical approach of the last section has practical and educational value, it is also valuable to consider the structure of the flow in more detail, and to examine how higher level solutions to the flow might be used to predict the performance of a cascade of a particular geometry. In doing so, it is important to distinguish between performance characteristics that are the result of idealized inviscid flow and those that are caused by viscous effects. Consider, first, the inviscid flow effects. König (1922) was the first to solve the potential flow through a linear cascade, in particular for a simple cascade of infinitely thin, straight blades. The solution leads to values of the deviation, δ, that, in turn, allow evaluation of the shut-off head coefficient, ψ_0, through equation 3.17. This is shown as a function of solidity in figure 3.4. Note that for solidities greater than about unity, the idealized, potential flow exits the blade passages parallel to the blades, and hence $\psi_0 \to 1$.

Another approach to the same issue of relating the flow angle, β_2, to the blade angles, is to employ an empirical rule for the deviation angle, $\delta = \beta_{b2} - \beta_2$ (equation 2.2), in terms of other geometric properties of the cascade. One early empirical relation suggested by Constant (1939) (see Horlock 1973) relates the deviation to the camber angle, θ_c, and the solidity, s, through

$$\delta_N = C \, \theta_c / s^{\frac{1}{2}} \tag{3.19}$$

where the subscript N refers to nominal conditions, somewhat arbitrarily defined as the operating condition at which the deflection $(\beta_2 - \beta_1)$ has a value that is 80% of that at which stall would occur. Constant suggested a value of 0.26 for the constant, C. Note that β_2 can then be evaluated and the head rise obtained from the characteristic 3.12. Later investigators explored the variations in the deviation angle with other flow parameters (see, for example, Howell 1942), and devised more complex correlations for use in the design of axial flow rotors (Horlock 1973). However, the basic studies of Leiblein on the boundary layers in linear cascades, and the role which these viscous effects play in determining the deviation and the losses, superceded much of this empirical work.

3.4 Viscous Effects in Linear Cascades

It is also of value to examine in more detail the mechanism of viscous loss in a cascade. Even in two-dimensional cascade flow, the growth of the boundary layers on the pressure and suction surfaces of the blades, and the wakes they form downstream of the blades (see figure 3.5), are complex, and not amenable to simple analysis. However, as the reviews by Roudebush and Lieblein (1965) and Lieblein (1965) demonstrate, it is nevertheless possible to provide some qualitative guidelines for the

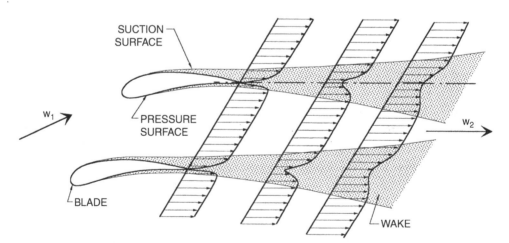

Figure 3.5. Sketch of the boundary layers on the surfaces of a cascade and the resulting blade wakes.

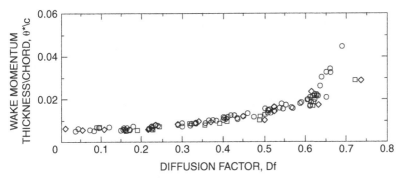

Figure 3.6. Correlation of the ratio of the momentum thickness of the blade wakes, θ^*, to the chord, c, with the diffusion factor, Df, for cascades of blades with three different profiles: NACA $65 - (A_{10})$ 10 series (○) and two British C.4 parabolic arc profiles (□ and ◇). The maximum thickness of the blades is $0.1c$ and the Reynolds number is 2.5×10^5. Adapted from Lieblein (1965).

resulting viscous effects on cascade performance. In this respect, the diffusion factor, introduced by Lieblein *et al.* (1953), is a useful concept that is based on the following approximations. First, we note that under normal operating conditions, the boundary layer on the suction surface will be much thicker than that on the pressure surface of the foil, so that, to a first approximation, we may neglect the latter. Then, the thickness of the wake (and therefore the total pressure loss) will be primarily determined by that fraction of the suction surface over which the velocity gradient is adverse, since that is where the majority of the boundary layer growth occurs. Therefore, Lieblein *et al.* argued, the momentum thickness of the wake, θ^*, should correlate with a parameter they termed the *diffusion factor*, given by $(w_{max} - w_2)/w_{max}$, where w_{max} is the maximum velocity on the suction surface. One should visualize deceleration or diffusion of the flow from w_{max} to w_2, and that this diffusion is the primary factor in determining the wake thickness. However, since w_{max} is not easily determined, Lieblein *et al.* suggest an approximation to the diffusion factor that is denoted Df, and given by

$$
\begin{aligned}
Df &= 1 - \frac{w_2}{w_1} + \frac{v_{\theta 2} - v_{\theta 1}}{2 s w_1} \\
&= 1 - \frac{\sin \beta_1}{\sin \beta_2} + \frac{\sin \beta_1 (\cot \beta_1 - \cot \beta_2)}{2s}
\end{aligned}
\tag{3.20}
$$

Figure 3.6 shows the correlation of the momentum thickness of the wake (normalized by the chord) with this diffusion factor, Df, for three foil profiles. Such correlations are now commonly used to determine the viscous loss due to blade boundary layers and wakes. Note that, once θ^*/c has been determined from such a correlation, the drag coefficient, C_D, and the friction or loss coefficient follow from equations 3.7,

3.9, and 3.10 and the fact that $D = \rho w_2^2 \theta^*$:

$$C_D = \frac{2\sin^2 \beta_M}{\sin^2 \beta_2} \frac{\theta^*}{c}; \quad f = \frac{s \sin \beta_M}{\sin^2 \beta_2} \frac{\theta^*}{c} \qquad (3.21)$$

The data shown in figure 3.6 were for a specific Reynolds number, Re, and the correlations must, therefore, be supplemented by a statement on the variation of the loss coefficient with Re. A number of correlations of this type exist (Roudebush and Lieblein 1965), and exhibit the expected decrease in the loss coefficient with increasing Re. For more detail on viscous losses in a cascade, the reader should consult the aforementioned papers by Lieblein.

In an actual turbomachine, there are several additional viscous loss mechanisms that were not included in the cascade analyses discussed above. Most obviously, there are additional viscous layers on the inner and outer surfaces that bound the flow, the hub and the shroud (or casing). These often give rise to complex, three-dimensional secondary flows that lead to additional viscous losses (Horlock and Lakshminarayana 1973). Moreover, the rotation of other, "non-active" surfaces of the impeller will lead to viscous shear stresses, and thence to losses known as "disk friction losses" in the terminology of turbomachines. Also, leakage flows from the discharge back to the suction, or from one stage back to a preceding stage in a multistage pump, constitute effective losses that must be included in any realistic evaluation of the losses in an actual turbomachine (Balje 1981).

3.5 Radial Cascade Analyses

Two-dimensional models for centrifugal or radial turbomachines begin with analyses of the flow in a radial cascade (section 2.2 and figure 3.7), the counterpart of the linear cascade for axial flow machines. More specifically, the counterpart of the linear flat plate cascade is the logarithmic spiral cascade, defined in section 2.2, and shown in more detail in figure 3.7. There exist simple conformal mappings that allow potential flow solutions for the linear cascade to be converted into solutions for the corresponding radial cascade flow, though the proper interpretation of these solutions requires special care. The resulting head/flow characteristic for frictionless flow in a radial cascade of infinitely thin logarithmic spiral blades is given in a classic paper by Busemann (1928), and takes the form

$$\psi = Sf_B - \psi_0 \phi \left(\cot \beta_b + \frac{v_{\theta 1}}{v_{m1}} \right) \qquad (3.22)$$

The terms Sf_B and ψ_0 result from quite separate and distinct fluid mechanical effects. The term involving ψ_0 is a consequence of the frictionless, potential flow head rise through any simple, nonrotating cascade whether of axial, radial, or mixed flow geometry. Therefore, ψ_0 is identical to the quantity, ψ_0, defined by equation 3.17 in

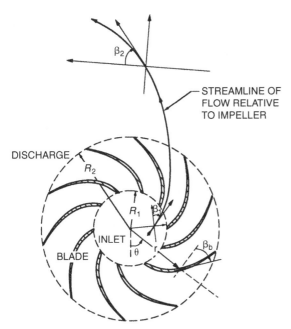

Figure 3.7. Schematic of the radial cascade corresponding to the linear cascade of figure 3.1.

the context of a linear cascade. The values for ψ_0 for a simple cascade of infinitely thin blades, whether linear, radial or mixed flow, are given in figure 3.4. The ψ_0 term can be thought of as the "through flow" effect, and, as demonstrated by figure 3.4, the value of ψ_0 rapidly approaches unity when the solidity increases to a value a little greater than one.

However, it is important to recognize that the ψ_0 term is the result of a frictionless, potential flow solution in which the vorticity is zero. This solution would be directly applicable to a static or nonrotating radial cascade in which the flow entering the cacade has no component of the vorticity vector in the axial direction. This would be the case for a nonswirling axial flow that is deflected to enter a nonrotating, radial cascade in which the axial velocity is zero. But, relative to a *rotating* radial cascade (or centrifugal pump impeller), such an inlet flow does have vorticity, specifically a vorticity with magnitude 2Ω and a direction of rotation opposite to the direction of rotation of the impeller. Consequently, the frictionless flow through the impeller is not irrotational, but has a constant and uniform vorticity of -2Ω.

In inviscid fluid mechanics, one frequently obtains solutions for these kinds of rotational flows in the following way. First, one obtains the solution for the irrotational flow, which is represented by ψ_0 in the current problem. Mathematically, this is the complementary solution. Then one adds to this a particular solution that satisfies all the same boundary conditions, but has a uniform vorticity, -2Ω. In the present context, this particular, or rotational, solution leads to the term, Sf_B, which, therefore, has a

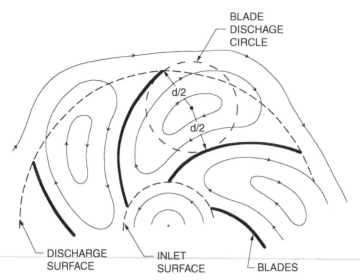

Figure 3.8. A sketch of the displacement component of the inviscid flow through a rotating radial cascade.

quite different origin from the irrotational term, ψ_0. The division into the rotational solution and the irrotational solution is such that all the net volumetric flow through the impeller is included in the irrotational (or ψ_0) component. The rotational solution has no through flow, but simply consists of a rotation of the fluid within each blade passage, as sketched in figure 3.8. Busemann (1928) called this the *displacement* flow; other authors refer to its rotating cells as *relative eddies* (Balje 1980, Dixon 1978). In his pioneering work on the fluid mechanics of turbomachines, Stodola (1927) was among the first to recognize the importance of this rotational component of the solution. Busemann (1928) first calculated its effect upon the head/flow characteristic for the case of infinitely thin, logarithmic spiral blades, in other words the simple cascade in the radial configuration. For reasons which will become clear shortly, the function, Sf_B, is known as the Busemann slip factor, and Busemann's solutions lead to the values presented in figure 3.9 when the solidity, $s > 1.1$. Note that the values of Sf_B are invariably less than or equal to unity, and, therefore, the effect of the displacement flow is to cause a decrease in the head. This deficiency can, however, be minimized by using a large number of blades. As the number of blades gets larger, Sf_B tends to unity as the rotational flow within an individual blade passage increasingly weakens. In practice, however, the frictional losses will increase with the number of blades. Consequently, there is an important compromise that must be made in choosing the number of blades. As figure 3.9 shows, this compromise will depend on the blade angle. Furthermore, the compromise must also take into account the structural requirements for the blades. Thus, radial machines for use with liquids usually have a smaller number of blades than those used for gases. The reason for this is that a liquid turbomachine

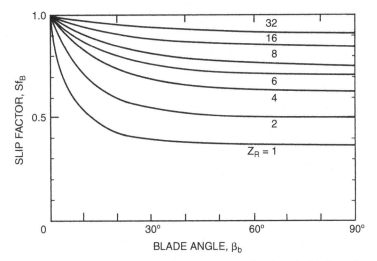

Figure 3.9. The Busemann slip factor, Sf_B, plotted against the blade angle, β_b, for various numbers of blades, Z_R. The results shown are for radial cascades of infinitely thin logarithmic spiral blades with solidities, $s > 1.1$. Adapted by Sabersky, Acosta and Hauptmann (1989) and Wislicenus (1947) from Busemann's (1928) theory.

requires much thicker blades, and, therefore, each blade creates much more flow blockage than in the case of a gas turbomachine. Consequently, liquid machines tend to have a smaller number of blades, typically eight for the range of specific speeds for which radial machines are designed ($N_D < 1.5$) (Stepanoff 1948, Anderson). Another popular engineering criterion (Stepanoff 1948) is that Z_R should be one third of the discharge blade angle, β_b (in degrees).

The decrease in the head induced by the displacement flow is due to the nonuniformity in the discharge flow; this nonuniformity results in a *mean* angle of discharge (denoted by β_2) that is different from the discharge blade angle, β_{b2}, and, therefore, implies an effective deviation angle or *slip*, Sf (see section 2.1). In fact, it is clear that the relations 2.16, 2.32, 3.22, and 2.4 imply that $Sf = Sf_B$, and, hence, the terminology used above. Stodola (1927) recognized that slip would be a consequence of the displacement flow, and estimated the magnitude of the slip velocity, $v_{\theta s}$, in the following approximate way. He argued that the slip velocity could be roughly estimated as $\Omega d/2$, where $d/2$ is the radius of the *blade discharge circle* shown in figure 3.8. He visualized this as representative of the rotating cell of fluid in a blade passage, and that the rotation of this cell at Ω would lead to the aforementioned $v_{\theta s}$. Then, provide Z_R is not too small, $d \approx 2\pi R_2 \sin \beta_{b2}$, and it follows that

$$v_{\theta s} = \pi \Omega R_2 \sin \beta_{b2}/Z_R \tag{3.23}$$

and, from equation 2.4, that the estimated slip factor, Sf_S, is

$$Sf_S = 1 - \frac{\pi \sin \beta_{b2}}{Z_R} \tag{3.24}$$

Numerical comparisons with the more exact results of Busemann presented in figure 3.9, show that equation 3.24 gives a reasonable first approximation. For example, an impeller with four blades, a blade angle of $25°$, and a solidity greater than unity, has a Stodola slip factor of $Sf_S = 0.668$ compared to the value of $Sf_B = 0.712$ from Busemann's more exact theory.

There is a substantial literature on slip factors for centrifugal pumps. Some of this focuses on the calculation of slip factors for inviscid flow in radial cascades with blades that are more complex than the infinitely thin, logarithmic spiral blades used by Busemann. Useful reviews of some of this work can be found, for example, in the work of Wislicenus (1947), Stanitz (1952), and Ferguson (1963). Other researchers attempt to find slip factors that provide the best fit to experimental data. In doing so, they also attempt to account for viscous effects in addition to the inviscid effect for which the slip factor was originally devised. As an example of this approach, the reader may consult Wiesner (1967), who reviews the existing, empirical slip factors, and suggests one that seems to yield the best comparison with the experimental measurements.

3.6 Viscous Effects in Radial Flows

We now turn to a discussion of the viscous effects in centrifugal pumps. Clearly a radial cascade will experience viscous boundary layers on the blades that are similar to those discussed earlier for axial flow machines (see section 3.4). However, two complicating factors tend to generate loss mechanisms that are considerably more complicated. These two factors are flow separation and secondary flow.

Normally, the flow in a centrifugal pump separates from the suction surface near the leading edge, and produces a substantial wake on the suction surfaces of each of the blades. Fischer and Thoma (1932) first identified this phenomenon, and observed that the wake can occur even at design flow. Normally, it extends all the way to the impeller discharge. Consequently, the discharge flow consists of a low velocity zone or wake next to the suction surface, and, necessarily, a flow of increased velocity in the rest of the blade passage. This "jet-wake structure" of the discharge is sketched in figure 3.10. Note that this viscous effect tends to counteract the displacement flow of figure 3.8. Since the work of Fischer and Thoma, many others have studied this aspect of flows in centrifugal pumps and compressors (see, for example, Acosta and Bowerman 1957, Johnston and Dean 1966, Eckardt 1976), and it is now recognized as essential to take these features into account in constructing any model of the flow in radial turbomachines. Modern analyses of the flow in radial turbomachines usually incorporate the basic features of the jet-wake structure in the blade passages (for example, Sturge and Cumpsty 1975, Howard and Osborne 1977). Sturge and Cumpsty have calculated the shape of the wake in a typical, two-dimensional radial cascade, using numerical methods to solve a free streamline problem similar to those discussed in chapter 7.

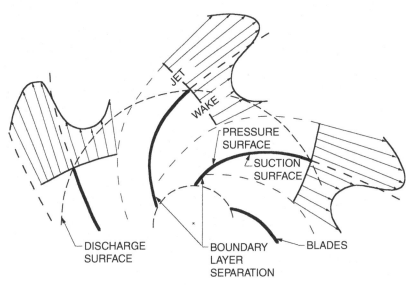

Figure 3.10. A sketch of actual discharge flow from a centrifugal pump or compressor including the alternating pattern of jets and wakes resulting from flow separation from the suction surfaces.

At design flow, the wake or boundary layer on the suction surface may be quite thin, but as the flow coefficient, ϕ, is decreased, the increased incidence leads to larger wakes (Fischer and Thoma 1932, Johnston and Dean 1966). Clearly, the nonuniformity of the discharge flow implies an "effective" slip due to these viscous effects. This slip will not only depend on the geometry of the blades but will also be a function of the flow coefficient and the Reynolds number. The change with flow coefficient is particularly interesting. As ϕ is decreased below the design value and the wake grows in width, an increasing fraction of the flow is concentrated in the jet. Johnston and Dean (1966) showed that this results in a flow that *more* closely follows the geometry of the pressure surface, and, therefore, to a *decrease* in the slip. This can be a major effect in radial compressors. Johnston and Dean made measurements in an 18-bladed radial compressor impeller with a 90° discharge blade angle (for which $Sf_S = 0.825$), and found that the effective slip factor increased monotonically from a value of about 0.8 at $\phi_2 = 0.5$ to a value of 1.0 at $\phi_2 = 0.15$. However, this increase in the slip factor did not produce an increase in the head rise, because the increase in the viscous losses was greater than the potential gain from the decrease in the slip.

Finally, it is important to recognize that secondary flows can also have a substantial effect on the development of the blade wakes, and, therefore, on the jet-wake structure. Moreover, the geometric differences between the typical radial compressor and the typical centrifugal pump can lead to significant differences in the secondary flows, the loss mechanisms, and the jet-wake structure. The typical centrifugal pump geometry was illustrated in figure 2.7, to which we should append the typical number of blades, $Z_R = 8$. A typical example is the geometry at $N_D = 0.6$, namely $R_{T1}/R_{T2} \approx 0.5$ and $B_2 \approx 0.2 R_{T2}$. Assuming $Z_R = 8$ and a typical blade angle at discharge of 25°, it follows that the blade passage flow at discharge has cross-sectional dimensions normal to the relative velocity vector of $0.2R_{T2} \times 0.3R_{T2}$, while the length of the blade passage is approximately $1.2R_{T2}$. Thus the blade passage is fairly wide relative to its length. In contrast, the typical radial compressor has a much smaller value of B_2/R_{T2}, and a much larger number of blades. As a result, not only is the blade passage much narrower relative to its length, but also the typical cross-section of the discharge flow is far from square, being significantly narrower in the axial direction. The viscous boundary layers on the suction and pressure surfaces of the blades, and on the hub and shroud (or casing), will have a greater effect the smaller the cross-sectional dimensions of the blade passage are relative to its length. Moreover, the secondary flows that occur in the corners of this passage amplify these viscous effects. Consequently, the flow that discharges from a blade passage of a typical radial compressor is more radically altered by these viscous effects than the flow discharging from a typical centrifugal pump.

4

Other Flow Features

4.1 Introduction

In this chapter we briefly survey some of the other important features of the flows through turbomachines. We begin with a section on the three-dimensional characteristics of flows, and a discussion of some of the difficulties encountered in adapting the cascade analyses of the last chapter to the complex geometry of most turbomachines.

4.2 Three-Dimensional Flow Effects

The preceding chapter included a description of some of the characteristics of two-dimensional cascade flows in both the axial and radial geometries. It was assumed that the flow in the meridional plane was essentially two-dimensional, and that the effects of the velocities (and the gradients in the velocity or pressure) normal to the meridional surface were neglible. Moreover, it was tacitly assumed that the flow in a real turbomachine could be synthesized using a series of these two-dimensional solutions for each meridional annulus. In doing so it is implicitly assumed that each annulus corresponds to a streamtube such as depicted in figure 4.1 and that the geometric relations between the inlet location, r_1, and thickness, dr_1, and the discharge thickness, dn, and location, r_2, are known *a priori*. In practice this is not the case and quasi-three-dimensional methods have been developed in order to determine the geometrical relation, $r_2(r_1)$. These methods continue to assume that the streamsurfaces are axisymmetric, and, therefore, neglect the more complicated three-dimensional aspects of the flow exemplified by the secondary flows discussed below (section 4.6). Nevertheless, these methods allow the calculation of useful turbomachine performance characteristics, particularly under circumstances in which the complex secondary flows are of less importance, such as close to the design condition. When the turbomachine is operating far from the design condition, the flow within a blade passage may have streamsurfaces that are far from axisymmetric.

In the context of axial flow machines, several approximate methods have been employed in order to determine $r_2(r_1)$ as a part of a quasi-three-dimensional solution

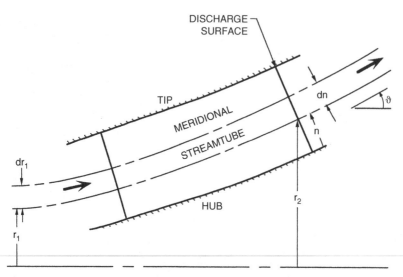

Figure 4.1. Geometry of a meridional streamtube in a pump impeller.

to the flow. Most of these are based on some application of the condition of radial equilibrium. In its simplest form, the radial equilibrium condition assumes that all of the terms in the equation of motion normal to the axisymmetric streamsurface are negligible, except for the pressure gradient and the centrifugal acceleration terms, so that

$$\frac{1}{\rho}\frac{dp}{dr} = \frac{v_\theta^2}{r} \qquad (4.1)$$

(The equivalent assumption in a radial machine would be that the axial pressure gradient is zero.) This assumption is differently embedded in several approaches to the solution of the flow. All of these use a condition like equation 4.1 (or some more accurate version) to relate the pressures in the different streamtubes upstream of the rotor (or stator), and a similar condition to connect the pressures in the streamtubes downstream of the rotor (or stator). When these relations are combined with the normal continuity and energy equations for each streamtube (that connect the conditions upstream with those at the downstream location), a complete set of equations is generated, and a solution to the flow can be obtained. In this class of meridional streamtube methods, the velocities normal to meridional streamsurfaces are largely neglected, but the cross-sectional areas of the streamtubes are adjusted to satisfy a condition based on the equation of motion normal to the meridional surface. Notable examples of this class of quasi-three-dimensional solutions are those devised at NASA Lewis by Katsanis and his co-workers (see Stockman and Kramer 1963, Katsanis 1964, Katsanis and McNally 1977).

The following example will illustrate one use of the "radial equilibrium" condition. We shall assume that the inlet flow is in radial equilibrium. This inlet flow is then divided into axisymmetric streamtubes, each with a specific radial location, r_1. Some initial estimate is made of the radial location of each of the streamtubes at discharge (in other words an estimate of the function $r_2(r_1)$). Then an iterative numerical method is employed, in which the total pressure rise through each streamtube is evaluated. Hence, the pressure distribution at discharge can be obtained. Then the width of each tube at discharge is adjusted ($r_2(r_1)$ is adjusted) in order to obtain the required radial pressure gradient between each pair of adjacent streamtubes. Subsequently, the process is repeated until a converged solution is reached. In some simple cases, analytical rather than numerical results can be obtained; an example is given in the next section.

More generally, it should be noted that quasi-three-dimensional analyses of this kind are often used for the design of axial turbomachines. A common objective is to achieve a design in which the total pressure is increasing (or decreasing) with axial position at the same rate at all radii, and, therefore, should be invariant with radial position. Combining this with the condition for radial equilibrium, leads to

$$\frac{d}{dr}\left(v_m^2\right) + \frac{1}{r^2}\frac{d}{dr}\left(r^2 v_\theta^2\right) = 0 \qquad (4.2)$$

If, in addition, we stipulate that the axial velocity, v_m, must be constant with radius, then equation 4.2 implies that the circumferential velocity, v_θ, must vary like $1/r$. Such an objective is termed a "free vortex" design. Another basic approach is the "forced vortex" design in which the circumferential velocity, v_θ, is proportional to the radius, r; then, according to the above equations, the axial velocity must decrease with r. More general designs in which $v_\theta = ar + b/r$ (a and b being constants) are utilized in practice for the design of axial compressors and turbines, with the objective of producing relatively uniform head rise and velocity at different radii (Horlock 1973). However, in the context of pumps, most of the designs are of the "forced vortex" type; Stepanoff (1948) lists a number of reasons for this historical development. Note that a forced vortex design with a uniform axial velocity would imply helical blades satisfying equation 2.7; thus many pumps have radial distributions of blade angle close to the form of that equation.

Radial equilibrium of the discharge flow may be an accurate assumption in some machines but not in others. When the blade passage is narrow (in both directions) relative to its length, the flow has adequate opportunity to adjust within the impeller or rotor passage, and the condition of radial equilibrium at discharge is usually reasonable. This is approximately the case in all pumps except propeller pumps of low solidity. However, in many compressors and turbines, the blade height is large compared with the chord and a radial equilibrium assumption at discharge is not appropriate. Under these circumstances, a very different approach utilizing an "actuator disc" has been successfully employed. The flows far upstream and downstream of the blade row are assumed to be in radial equilibrium, and the focus is on the adjustment of the flow between these locations and the blade row (see figure 4.2). The flow through the blade

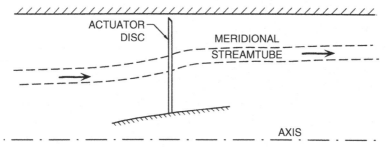

Figure 4.2. Actuator disc model of an axial blade row with a generic meridional streamtube.

row itself is assumed to be so short that the streamsurfaces emerge at the same radial locations at which they entered; thus the blade row is modeled by an infinitesmally thin "actuator disc." In some respects, the actuator disc approach is the opposite of the radial equilibrium method; in the former, all the streamline adjustment is assumed to occur external to the blade passages whereas, in many radial equilibrium applications, the adjustment all occurs internally.

Since actuator disc methods are rarely applied in the context of pumps we shall not extend the discussion of them further. More detail can be found in texts such as Horlock (1973). We shall, however, provide an example of a radial equilibrium analysis since the results will prove useful in a later chapter.

4.3 Radial Equilibrium Solution: An Example

For the purposes of this example of a radial equilibrium solution, the flow through the pump impeller is subdivided into streamtubes, as shown in figure 4.1. We choose to examine one generic streamtube with an inlet radius, r_1, and thickness, dr_1. Both the position, n, and the thickness, dn, of the streamtube at discharge are not known *a priori*, and must be determined as a part of the solution. Conservation of mass requires that

$$v_{m1} r_1 dr_1 = v_{m2}(n)(R_{H2} + n\cos\vartheta)dn \qquad (4.3)$$

where n is a coordinate measured normal to the streamlines at discharge and $n = 0$ at the hub so that $r_2 = R_{H2} + n\cos\vartheta$.

Applying the radial equilibrium assumption, the pressure distribution over the exit plane is given by

$$\frac{1}{\rho}\frac{\partial p_2}{\partial n} = \frac{v_{\theta 2}^2 \cos\vartheta}{R_{H2} + n\cos\vartheta} \qquad (4.4)$$

It is also necessary to specify the variation of the discharge blade angle, $\beta_{b2}(n)$, with position, and, for the reasons described in section 4.2, we choose the helical distribution given by equation 2.7. Note that this implies helical blades in the case of an axial flow pump with $\vartheta = 0$, and a constant β_{b2} in the case of a centrifugal pump with $\vartheta = 90°$. Moreover, we shall assume that the flow at discharge is parallel with the blades so that $\beta_2(n) = \beta_{b2}(n)$.

The formulation of the problem is now complete, and it is a relatively straightforward matter to eliminate $p_2(n)$ from equations 2.30 and 4.4, and then use the velocity triangles and the continuity equation 4.3 to develop a single differential equation for $v_{m2}(n)$. Assuming that the inlet is free of swirl, and that v_{m1} is a constant, this equation for $v_{m2}(n)$ can then be integrated to obtain the velocity and pressure distributions over the exit. It remains to evaluate the total energy added to the flow by summing the energies added to each of the streamtubes according to equation 2.21:

$$H = \frac{1}{Q} \int_{HUB}^{TIP} \frac{(p_2^T - p_1^T)}{\rho g} 2\pi r_2 v_{m2} dn \tag{4.5}$$

Nondimensionalizing the result, we finally obtain the following analytical expression for the performance:

$$\psi = \Sigma_1 + \Sigma_2 \phi_2 + \Sigma_3/\phi_2 \tag{4.6}$$

where Σ_1, Σ_2, and Σ_3 are geometric quantities defined by

$$\Sigma_2 = \frac{\Gamma \cot \beta_{bT2}}{\ell n \Gamma^*} \left[1 + \frac{\Gamma \sin^2 \beta_{bT2} \cos^2 \beta_{bT2}}{\Gamma^* \ell n \Gamma^*} \right]$$

$$\Sigma_3 = \tan^3 \beta_{bT2} \left[1 - \frac{\Gamma^2 \cos^4 \beta_{bT2}}{\Gamma^* \{\ell n \Gamma^*\}^2} \right] \tag{4.7}$$

$$\Sigma_1 = -\Sigma_3 \cot \beta_{bT2} - \Sigma_2 \tan \beta_{bT2}$$

where Γ and Γ^* are given by

$$\Gamma = 1 - \left(\frac{R_{H2}}{R_{T2}} \right)^2; \quad \Gamma^* = 1 - \Gamma \cos^2 \beta_{bT2} \tag{4.8}$$

Thus the geometric quantities, Σ_1, Σ_2, and Σ_3, are functions only of Γ and β_{bT2}.

Examples of these analytical performance curves are given later in figures 7.13 and 7.15. Note that this idealized hydraulic performance is a function only of the geometric variables, Γ and β_{bT2}, of the discharge. Moreover, it is readily shown that in the centrifugal limit of $\Gamma \to 0$ then $\Sigma_1 \to 1$, $\Sigma_2 \to -\cot \beta_{bT2}$, $\Sigma_3 \to 0$, and the earlier result of equation 2.33 is recovered.

It is of interest to explore some optimizations based on the hydraulic performance, given by equation 4.6. Though the arguments presented here are quite heuristic, the results are interesting. We begin with the observation that two particular geometric factors are important in determining the viscous losses in many internal flows.

If the cross-sectional area of the flow increases at more than a marginal rate, the deceleration-induced boundary layer separation and turbulence can lead to large viscous losses that might not otherwise occur. Consequently, the mean value of w_2/w_1 is an important design parameter, as implied earlier in section 3.4. In the present analysis, the mean value of this parameter is given by the area ratio, Ar^*, where, from geometric considerations,

$$Ar^* = \frac{\Gamma \sin \beta_{bT2}}{\cos \vartheta \, (R_{T1}/R_{T2})^2 \sin \left\{ \tan^{-1} (\phi_2 Ar R_{T2}/R_{T1}) \right\}} \tag{4.9}$$

We shall also use the ratio, Ar, of the area of the axisymmetric discharge surface to the area of the inlet surface given by

$$Ar = \Gamma/\cos \vartheta \, (R_{T1}/R_{T2})^2 \tag{4.10}$$

In this example it is assumed that $R_{H1} = 0$; non-zero values can readily be accommodated, but do not alter the qualitative nature of the results obtained.

Many centrifugal pumps are designed with Ar^* values somewhat greater than unity because the flow must subsequently be decelerated in the diffuser and volute, and smaller values of Ar^* would imply larger diffusion losses in those nonrotating components. But, from the point of view of minimizing losses in the impeller alone, one justifiable optimization would require $Ar^* \approx 1$.

The second geometric factor that can influence the magnitude of the viscous losses in an internal flow is the amount of turning imposed on the flow. In the present analysis, we shall make use of an angle, ϵ, describing the "angle of turn" of the flow as it proceeds through the turbomachine. It is defined as the angle of the discharge relative velocity vector to the conical discharge surface minus the angle of the inlet relative velocity vector to the inlet surface:

$$\epsilon = \beta_{bT2} - \tan^{-1} \{\phi_2 A_r R_{T2}/R_{T1}\} \tag{4.11}$$

Note that, in purely axial flow, the angle of turn, ϵ, is zero for the case of a flow with zero incidence through a set of helical blades of constant pitch. Also note that, in purely radial flow, the angle of turn, ϵ, is zero for the case of a flow with zero incidence through a set of logarithmic spiral blades. Therefore, using somewhat heuristic interpolation, one might argue that ϵ may be useful in the general case to describe the degree of turning applied to the flow by a combination of a nonzero incidence at inlet and the curvature of the blade passages.

For the purposes of this example, we now postulate that the major hydraulic losses encountered in the flow through the pump are minimized when ϵ is minimized. Let us assume that this minimum value of ϵ can be approximated by zero. Referring to this maximum efficiency point of operation as the "design point" (where conditions are denoted by the suffix, D), it follows from equation 4.11 that

$$\phi_{2D} = \frac{R_{T1}}{R_{T2}} \frac{\tan \beta_{bT2}}{Ar} \tag{4.12}$$

and hence that

$$\psi_D = \Sigma_1 + \Sigma_2 \phi_{2D} + \Sigma_3 / \phi_{2D} \tag{4.13}$$

Thus the specific speed for which the pump is designed, N_D, is given by

$$N_D = \left[\frac{\pi \tan \beta_{bT2} \left(1 - \frac{R_{H1}^2}{R_{T1}^2} \right)}{\left\{ \Sigma_1 \left(\frac{R_{T2}}{R_{T1}} \right)^2 + \frac{\Sigma_2 \tan \beta_{bT2}}{Ar} \left(\frac{R_{T2}}{R_{T1}} \right) + \frac{\Sigma_3 Ar}{\tan \beta_{bT2}} \left(\frac{R_{T2}}{R_{T1}} \right)^3 \right\}^{\frac{3}{2}}} \right]^{\frac{1}{2}} \tag{4.14}$$

and is a function only of the geometric quantities R_{T1}/R_{T2}, R_{H1}/R_{T1}, R_{H2}/R_{T2}, ϑ, and β_{bT2}.

Examine now the variation of N_D with these geometric variables, as manifest by equation 4.14, bearing in mind that the practical design problem involves the reverse procedure of choosing the geometry best suited to a known specific speed. The number of geometric variables will be reduced to four by assuming $R_{H1} = 0$. Note also that, at the design point given by equation 4.12, it follows that $Ar^* = Ar$ and it is more convenient to use this area ratio in place of the variable R_{H2}/R_{T2}. Thus we consider the variations of N_D with $\vartheta, \beta_{bT2}, R_{T1}/R_{T2}$, and Ar^*.

Calculations of N_D from equation 4.14 show that, for specific speeds less than unity, for sensible values of Ar^* of the order of unity, and for blade angles β_{bT2} which are less than about 70° (which is the case in well-designed pumps), the results are virtually independent of the angle ϑ, a feature that simplifies the parametric variations in the results. For convenience, we choose an arbitrary value of $\vartheta = 50°$. Then typical results for $Ar^* = 1.0$ are presented in figure 4.3, which shows the "optimum" R_{T1}/R_{T2} for various design specific speeds, N_D, at various discharge blade angles, β_{bT2}. Considering the heuristic nature of some of the assumptions that were used in this optimization, the agreement between the results and the conventional recommendation (reproduced from figure 2.7) is remarkable. It suggests that the evolution of pump designs has been driven by processes minimizing the viscous losses, and that this minimization involves the optimization of some simple geometric variables. The values of ψ_D and ϕ_D, that correspond to the results of figure 4.3, are plotted in figure 4.4. Again, the comparison of the traditional expectation and the present analysis is good, except perhaps at low specific speeds where the discrepancy may be due to the large values of Ar^* which are used in practice. Finally, we observe that one can construct sets of curves, such as those of figure 4.3, for other values of the area ratio, Ar^*. However, for reasonable values

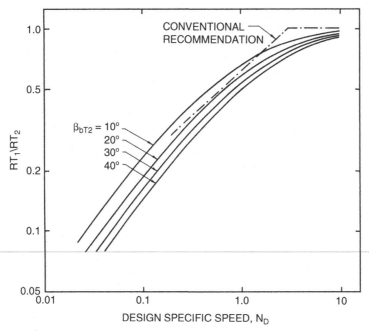

Figure 4.3. Comparison of the results of equation 4.14 with the conventional recommendation from figure 2.7 for the optimum ratio of inlet to discharge tip radius as a function of design specific speed, N_D.

of β_{bT2} like 20°, the curves for $0.8 < Ar^* < 2.0$ do not differ greatly from those for $Ar^* = 1.0$.

The foregoing analysis is intended only as an example of the application of the radial equilibrium methodology, and the postscript is included because of the interesting results it produces. Clearly some of the assumptions in the postscript are approximate, and would be inappropriate in any accurate analysis of the viscous losses.

4.4 Discharge Flow Management

To this point the entire focus has been on the flow within the impeller or rotor of the pump. However, the flow that discharges from the impeller requires careful handling in order to preserve the gains in energy imparted to the fluid. In many machines this requires the conversion of velocity head to pressure by means of a diffuser. This inevitably implies hydraulic losses, and considerable care needs to be taken to minimize these losses. The design of axial and radial diffusers, with and without vanes to recover the swirl velocity, is a major topic, whose details are beyond the scope of this book. The reader is referred to the treatise by Japikse (1984).

Such diffusers are more common in compressors than in pumps. Typical pump configurations are as follows. Axial flow pumps often employ a set of stator vanes

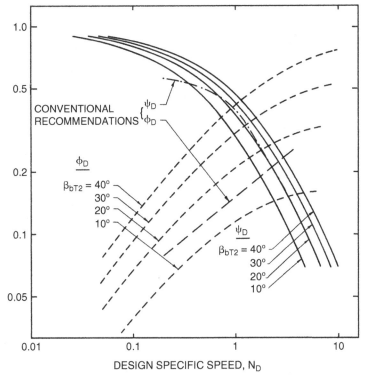

Figure 4.4. Comparison of the results of equation 4.14 with the conventional recommendation of figure 2.7 for the head coefficient, ψ_D, and the flow coefficient, ϕ_D, as functions of the design specific speed, N_D.

before (or in) the axial diffuser in order to recover the swirl velocities. Special care needs to be taken to match the swirl angles of the flow exiting the impeller with the inlet angles of the stator vanes. It is advisable, where possible, to measure the impeller discharge flow directly before finalizing a design. In some designs, the axial diffuser will be followed by a spiral collector or "volute" in order to recover the energy in the remaining swirl and axial velocities.

In the case of centrifugal pumps, a radial flow diffuser with vanes may or may not be used. Often it is not, and the flow discharges directly into the volute. The proper design of this volute is an important component of centrifugal pump design (Anderson 1955, Worster 1963, Stepanoff 1948). The objective is to design a volute in which the flow is carefully matched to the flow exiting the impeller, so that the losses are minimized and so that the pressure is uniform around the impeller discharge. The basic concept is sketched in figure 4.5. The flow discharges from the impeller with a velocity, $v_{\theta 2}$, in the tangential direction and a velocity, v_{r2}, in the radial direction, given by

$$v_{r2} = Q/2\pi R_2 B_2 \qquad (4.15)$$

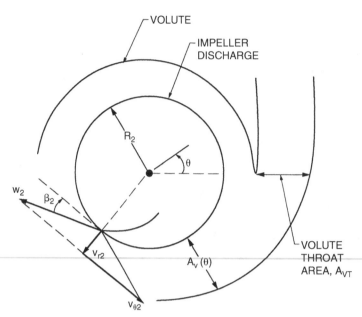

Figure 4.5. Volute flow notation.

where B_2 is the impeller discharge width. For simplicity, it will be assumed that the discharge from the impeller is circumferentially uniform; in fact, nonuniform volute flow will lead to a nonuniform flow in the impeller that is unsteady in the rotating frame. Though this complication is often important, it is omitted from the present, simple analysis.

If we further assume radially uniform velocity at each angular location in the volute (also an assumption that needs to be modified in a more accurate analysis), then it follows from the application of conservation of mass to an element, $d\theta$, of the volute that the discharge flow will be matched to the flow in the volute if

$$v_{\theta 2}\frac{dA_V}{d\theta} = v_{r2}R_2B_2 \qquad (4.16)$$

This requires a circumferentially uniform rate of increase of the volute area of $dA_V/d\theta = v_{r2}R_2B_2/v_{\theta 2}$ over the entire development of the spiral. If the area of the clearance between the cutwater and the impeller discharge is denoted by A_{VC}, and the volute exit area is denoted by A_{VT}, then A_V should have the following linear behavior:

$$A_V = A_{VC} + \frac{\theta}{2\pi}A_{VT} \qquad (4.17)$$

It follows that $dA_V/d\theta = A_{VT}/2\pi$ and hence

$$\phi^{-1}\tan\beta_2 - 1 = \frac{2\pi R_2 B_2 \tan\beta_2}{A_{VT}} \qquad (4.18)$$

Consequently, for a given impeller operating at a given design flow coefficient, ϕ_D, there exists a specific area ratio, $2\pi R_2 B_2 \tan \beta_2 / A_{VT}$, for the volute geometry. This parameter is close to the ratio which Anderson (1955) used in his design methodology (see also Worster 1963), namely the ratio of the cross-sectional area of the flow leaving the impeller ($2\pi R_2 B_2 \sin \beta_2$) to the volute throat area (A_{VT}). For more detailed analyses of the flow in a volute, the reader is referred to Pfleiderer (1932), Stepanoff (1948), and Lazarkiewicz and Troskolanski (1965). For example, Pfleiderer explored the radially nonuniform distributions of velocity within the volute and the consequences for the design methodology.

One of the other considerations during the design of a volute is the lateral force on the impeller that can develop due to circumferentially nonuniform flow and pressure in the volute. These, and other related issues, are discussed in chapter 10.

4.5 Prerotation

Perhaps no aspect of turbomachinery flow is more misrepresented and misunderstood than the phenomenon of "prerotation." While this belongs within the larger category of secondary flows (dealt with in section 4.6), it is appropriate to address the issue of prerotation seperately, not only because of its importance for the hydraulic performance, but also because of its interaction with cavitation.

It is first essential to distinguish between two separate phenomena both of which lead to a swirling flow entering the pump. These two phenomena have very different fluid mechanical origins. Here we shall distinguish them by the separate terms, "backflow-induced swirl" and "inlet prerotation." Both imply a swirl component of the flow entering the pump. In fluid mechanical terms, the flow has axial vorticity (if the axis of rotation is parallel with the axis of the inlet duct) with a magnitude equal to twice the rate of angular rotation of the swirl motion. Moreover, there are some basic properties of such swirling flows that are important to the understanding of prerotation. These are derived from the vorticity transport theorem (see, for example, Batchelor 1967). In the context of the steady flow in an inlet duct, this theorem tells us that the vorticity will only change with axial location for two reasons: (a) because vorticity is diffused into the flow by the action of viscosity, or (b) because the flow is accelerated or decelerated as a result of a change in the cross-sectional area of the flow. The second mechanism results in an increase in the swirl velocity due to the stretching of the vortex line, and is similar to the increase in rotation experienced by figure skaters when they draw their arms in closer to their body. When the moment of inertia is decreased, conservation of angular momentum results in an increase in the rotation rate. Thus, for example, a nozzle in the inlet line would increase the magnitude of any preexisting swirl.

For simplicity, however, we shall first consider inlet ducting of uniform and symmetric cross-sectional area, so that only the first mechanism exists. In inviscid flow, it follows that, if there is a location far upstream at which the swirl (or axial vorticity) is zero, then, in the absence of viscous effects, the swirl will be everywhere zero. This important result, which is a version of Kelvin's theorem (Batchelor 1967), is not widely recognized in discussions of prerotation. Moreover, the result is not altered by the existence of viscous effects, since purely axial motion cannot generate axial vorticity. However, there are two common circumstances in which prerotation can be generated without violation of the above theorem, and these give rise to the two phenomena named earlier.

The first of these common circumstances arises because of one of the most important secondary flows that can occur in pumps, namely the phenomenon of "backflow." This is caused by the leakage flow between the tip of the blades of an impeller (we consider first an unshrouded impeller) and the pump casing. The circumstances are depicted in figure 4.6. Below a certain critical flow coefficient, the pressure difference driving the leakage flow becomes sufficiently large that the tip leakage jet penetrates upstream of the inlet plane of the impeller, and thus forms an annular region of "backflow" in the inlet duct. After penetrating upstream a certain distance, the fluid of this jet is then entrained back into the main inlet flow. The upstream penetration distance increases with decreasing flow coefficient, and can reach many diameters upstream

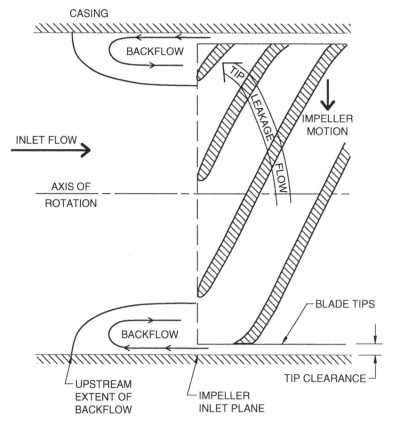

Figure 4.6. Lateral view of impeller inlet flow showing tip leakage flow leading to backflow.

of the inlet plane. In some pump development programs (such as the Rocketdyne J-2 liquid oxygen pump) efforts have been made to insert a "backflow deflector" in order to improve pump performance (Jakobsen 1971). The intention of such a device is to prevent the backflow from penetrating too far upstream, to reduce the distortion of the inlet flow field, and to recover, as far as is possible, the swirl energy in the backflow. More recently, a similar device was successfully employed in a centrifugal pump (Sloteman *et al.* 1984).

Some measurements of the axial and swirl velocities just upstream of an axial inducer are presented in figure 4.7. This data is taken from del Valle *et al.* (1992), though very similar velocity profiles have been reported by Badowski (1969, 1970) (see also Janigro and Ferrini 1973), and the overall features of the flow are similar whether the pump is shrouded or unshrouded, axial or centrifugal (see, for example, Stepanoff 1948, Okamura and Miyashiro 1978, Breugelmans and Sen 1982, Sloteman *et al.* 1984). Measurements are shown in figure 4.7 for two distances upstream of the inlet plane (half a radius and one radius upstream), and for a number of flow coefficients, ϕ. Note from the axial flow velocity profiles that, as the flow coefficient is decreased, the backflow reaches a half radius upstream at about $\phi \approx 0.066$, and one radius upstream at about $\phi \approx 0.063$. The size of the backflow region grows as ϕ is decreased. It is particularly remarkable that at $\phi \approx 0.05$, nearly 30% of the inlet area

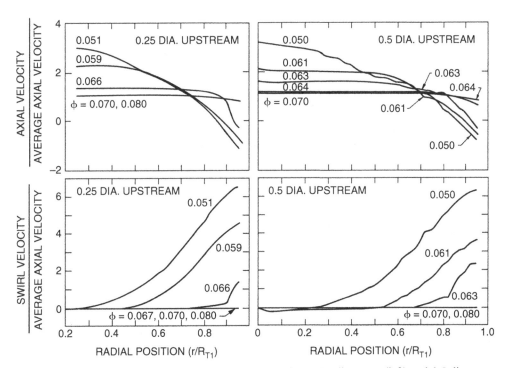

Figure 4.7. Axial and swirl velocity profiles in the inlet duct 0.25 diameters (left) and 0.5 diameters (right) upstream of the inlet plane of an inducer (Impeller VI) for various flow coefficients as shown (from del Valle, Braisted and Brennen 1992).

is experiencing reverse flow! We can further observe from the swirl velocity data that, in the absence of backflow, the inlet flow has zero swirl. Kelvin's theorem tells us this must be the case because the flow far upstream has no swirl.

Obviously the backflow has a high swirl velocity imparted to it by the impeller blades. But what is also remarkable is that this vorticity is rapidly spread to the core of the main inlet flow, so that at $\phi = 0.05$, for example, almost the entire inlet flow has a nonzero swirl velocity. The properties of swirling flows discussed above are not violated, since the origin of the vorticity is the pump itself and the vorticity is transmitted to the inflow via the backflow. The rapidity with which the swirl vorticity is diffused to the core of the incoming flow remains something of a mystery, for it is much too rapid to be caused by normal viscous diffusion (Braisted 1979). It seems likely that the inherent unsteadiness of the backflow (with a strong blade passing frequency component) creates extensive mixing which effects this rapid diffusion. However it arises, it is clear that this "backflow-induced swirl," or "prerotation," will clearly affect the incidence angles and, therefore, the performance of the pump.

Before leaving the subject of backflow, it is important to emphasize that this phenomenon also occurs at flow rates below design in centrifugal as well as axial flow pumps, and with shrouded as well as unshrouded impellers (see, for example, Okamura and Miyashiro 1978, Makay 1980). The detailed explanation may differ from one device to another, but the fundamental tendency for an impeller to exhibit this kind of secondary flow at larger angles of incidence seems to be universal.

But there is another, quite separate origin for prerotation, and this is usually manifest in practice when the fluid is being drawn into the pump from an "inlet bay" or reservoir with a free surface (figure 4.8). Under such circumstances, it is almost inevitable that the large scale flow in the reservoir has some nonuniformity that constitutes axial vorticity or circulation in the frame of reference of the pump inlet. Even though the fluid velocities associated with this nonuniformity may be very small, when the vortex lines are stretched as the flow enters the inlet duct, the vorticity is

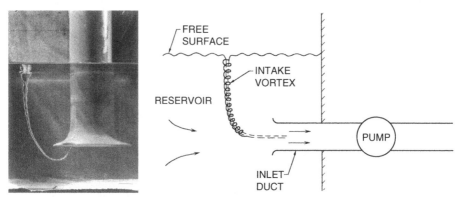

Figure 4.8. Right: sketch of a typical inlet vortex associated with prerotation. Left: Photograph of an air-filled inlet vortex from Wijdieks (1965) reproduced with permission of the Delft Hydraulics Laboratory.

greatly amplified, and the inlet flow assumes a significant preswirl or "inlet prerotation." The effect is very similar to the bathtub vortex. Once the flow has entered an inlet duct of constant cross-sectional area, the magnitude of the swirl usually remains fairly constant over the short lengths of inlet ducting commonly used.

Often, the existence of "inlet prerotation" can have unforeseen consequences for the suction performance of the pump. Frequently, as in the case of the bathtub vortex, the core of the vortex runs from the inlet duct to the free surface of the reservoir, as shown in figure 4.8. Due to the low pressure in the center of the vortex, air is drawn into the core and may even penetrate to the depth of the duct inlet, as illustrated by the photograph in figure 4.8 taken from the work of Wijdieks (1965). When this occurs, the pump inlet suddenly experiences a two-phase air/water flow rather than the single-phase liquid inlet flow expected. This can lead, not only to a significant reduction in the performance of the pump, but also to the vibration and unsteadiness that often accompany two-phase flow. Even without air entrainment, the pump performance is almost always deteriorated by these suction vortices. Indeed this is one of the prime suspects when the expected performance is not realized in a particular installation. These intake vortices are very similar to those which can occur in aircraft engines (De Siervi *et al.* 1982).

4.6 Other Secondary Flows

Most pumps operate at high Reynolds numbers, and, in this regime of flow, most of the hydraulic losses occur as a result of secondary flows and turbulent mixing. While a detailed analysis of secondary flows is beyond the scope of this monograph (the reader is referred to Horlock and Lakshminarayana (1973) for a review of the fundamentals), it is important to outline some of the more common secondary flows that occur in pumps. To do so, we choose to describe the secondary flows associated with three typical pump components, the unshrouded axial flow impeller or inducer, the shrouded centrifugal impeller, and the vaneless volute of a centrifugal pump.

Secondary flows in unshrouded axial flow inducers have been studied in detail by Lakshminarayana (1972, 1981), and figure 4.9, which was adapted from those publications, provides a summary of the kinds of secondary flows that occur within the blade passage of such an impeller. Dividing the cross-section into a core region, boundary layer regions on the pressure and suction surfaces of the blades, and an interference region next to the static casing, Lakshminarayana identifies the following departures from a simple flow following the blades. First, and perhaps most important, there will be a strong leakage flow (called the tip leakage or tip clearance flow) around the blade tips driven by the pressure difference between the pressure surface and the suction surface. Clearly this flow will become even more pronounced at flow rates below design when the blades are more heavily loaded. This leakage flow will entrain secondary flow on both surfaces of the blades, as shown by the dashed arrows in figure 4.9. Second, the flow in the boundary layers will tend to generate an

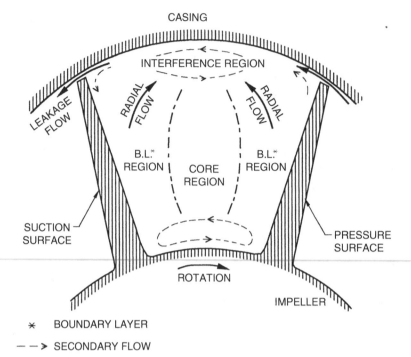

CASING

INTERFERENCE REGION

RADIAL FLOW

RADIAL FLOW

LEAKAGE FLOW

B.L.* REGION

CORE REGION

B.L.* REGION

SUCTION SURFACE

PRESSURE SURFACE

ROTATION

IMPELLER

∗   BOUNDARY LAYER

− − ➤ SECONDARY FLOW

Figure 4.9. Cross-section of a blade passage in an axial flow impeller showing the tip leakage flow, boundary layer radial flow, and other secondary flows (adapted from Lakshminarayana 1981).

outward radial component on both the suction and pressure surfaces, though the former may be stronger because of enhancement by the leakage flow. The photographs of figure 4.10, which are taken from Bhattacharyya *et al.* (1993), show a strong outward radial component of the flow on the blade surface of an inducer. This is particularly pronounced near the leading edge (left-hand photograph). Incidentally, Bhattacharyya *et al.* not only observed the backflow associated with the tip clearance flow, but also a "backflow" at the hub in which flow reenters the blade passage from downstream of the inducer. Evidence for this secondary flow can be seen on the hub surface in the right-hand photograph of figure 4.10. Finally, we should mention that Lakshminarayana also observed secondary vortices at both the hub and the casing as sketched in figure 4.9. The vortex near the hub was larger and more coherent, while a confused interference region existed near the casing.

Additional examples of secondary flows are given in the descriptions by Makay (1980) of typical flows through shrouded centrifugal impellers. Figure 4.11, which has been adapted from one of Makay's sketches, illustrates the kind of secondary flows that can occur at off-design conditions. Note, in particular, the backflow in the impeller eye of this shrouded impeller pump. This backflow may well interact in an important way with the discharge-to-suction leakage flow that is an important feature of the hydraulics of a centrifugal pump at all flow rates. As testament to the importance of the backflow, Makay cites a case in which the inlet guide vanes of a primary coolant

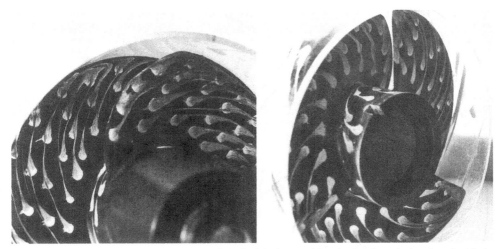

Figure 4.10. Photographs of a 10.2 *cm*, 12° helical inducer with a lucite shroud showing the blade surface flow revealed by the running paint dot technique. On the left the suction surfaces viewed from the direction of the inlet. On the right the view of the pressure surfaces and the hub from the discharge. The flow is for 2000 *rpm* and $\phi_1 = 0.041$. From Bhattacharyya *et al.* (1993).

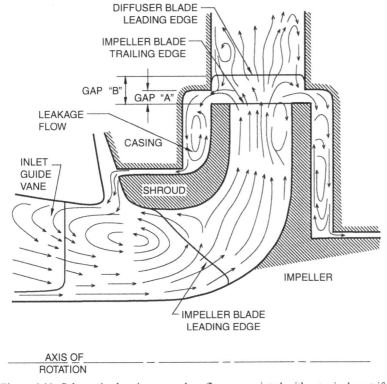

Figure 4.11. Schematic showing secondary flows associated with a typical centrifugal pump operating at off-design conditions (adapted from Makay 1980).

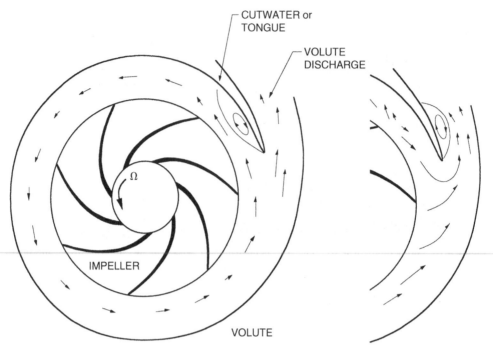

Figure 4.12. Schematic of a centrifugal pump with a single, vaneless volute indicating the disturbed and separated flows which can occur in the volute below (left) and above (right) the design flow rate.

pump in a power plant suffered structural damage due to the repeated unsteady loads caused by this backflow. Note should also be made of the secondary flows that Makay describes occurring in the vicinity of the impeller discharge.

   It is also important to mention the disturbed and separated flows that can often occur in the volute of a centrifugal pump when that combination is operated at off-design flow rates (Binder and Knapp 1936, Worster 1963, Lazarkiewicz and Troskolanski 1965, Johnston and Dean 1966). As described in the preceding section, and as indicated in figure 4.12, one of the commonest geometries is the spiral volute, designed to collect the flow discharging from an impeller in a way that would result in circumferentially uniform pressure and velocity. However, such a volute design is specific to a particular design flow coefficient. At flow rates above or below design, disturbed and separated flows can occur particularly in the vicinity of the cutwater or tongue. Some typical phenomena are sketched in figure 4.12 which shows separation on the inside and outside of the tongue at flow coefficients below and above design, respectively. It also indicates the flow reversal inside the tongue that can occur above design (Lazarkiewicz and Troskolanski 1965). Moreover, as Chu *et al.* (1993) have recently demonstrated, the unsteady shedding of vortices from the cutwater can be an important source of vibration and noise.

5

Cavitation Parameters and Inception

5.1 Introduction

This chapter will deal with the parameters that are used to describe cavitation, and the circumstances that govern its inception. In subsequent chapters, we address the deleterious effects of cavitation, namely cavitation damage, noise, the effect of cavitation on hydraulic performance, and cavitation-induced instabilities.

5.2 Cavitation Parameters

Cavitation is the process of the formation of vapor bubbles in low pressure regions within a flow. One might imagine that vapor bubbles are formed when the pressure in the liquid reaches the vapor pressure, p_V, of the liquid at the operating temperature. While many complicating factors discussed later cause deviations from this hypothesis, nevertheless it is useful to adopt this as a criterion for the purpose of our initial discussion. In practice, it can also provide a crude initial guideline.

The static pressure, p, in any flow is normally nondimensionalized as a pressure coefficient, C_p, defined as

$$C_p = (p - p_1)/\frac{1}{2}\rho U^2 \tag{5.1}$$

where p_1 is some reference static pressure for which we shall use the pump inlet pressure and U is some reference velocity for which we shall use the inlet tip speed, ΩR_{T1}. It is important to note that, for the flow of an incompressible liquid within rigid boundaries, C_p is only a function of the geometry of the boundaries and of the Reynolds number, Re, which, for present purposes, can be defined as $2\Omega R_{T1}^2/\nu$ where ν is the kinematic viscosity of the fluid. It is equally important to note that, in the absence of cavitation, the fluid velocities and the pressure coefficient are *independent* of the level of the pressure. Thus, for example, a change in the inlet pressure, p_1, will simply result in an equal change in all the other pressures, so that C_p is unaffected. It follows that, in any flow with prescribed fluid velocities, geometry and Reynolds number, there will be a particular location at which the pressure, p, is a minimum and that the difference between this minimum pressure, p_{min}, and the inlet pressure, p_1 is given by

$$C_{pmin} = (p_{min} - p_1)/\frac{1}{2}\rho U^2 \tag{5.2}$$

where C_{pmin} is some negative number which is a function only of the geometry of the device (pump) and the Reynolds number. If the value of C_{pmin} could be obtained either experimentally or theoretically, then we could establish the value of the inlet pressure, p_1, at which cavitation would first appear (assuming that this occurs when $p_{min} = p_V$) as p_1 is decreased, namely

$$(p_1)_{\substack{\text{CAVITATION} \\ \text{APPEARANCE}}} = p_V + \frac{1}{2}\rho U^2 \left(-C_{pmin}\right) \tag{5.3}$$

which for a given device, given fluid, and given fluid temperature, would be a function only of the velocity, U.

Traditionally, several special dimensionless parameters are utilized in evaluating the potential for cavitation. Perhaps the most fundamental of these is the cavitation number, σ, defined as

$$\sigma = (p_1 - p_V)/\frac{1}{2}\rho U^2 \tag{5.4}$$

Clearly every flow has a value of σ whether or not cavitation occurs. There is, however, a particular value of σ corresponding to the particular inlet pressure, p_1, at which cavitation first occurs as the pressure is decreased. This is called the cavitation inception number, and is denoted by σ_i:

$$\sigma_i = \left[(p_1)_{\substack{\text{CAVITATION} \\ \text{APPEARANCE}}} - p_V\right]/\frac{1}{2}\rho U^2 \tag{5.5}$$

If cavitation inception occurs when $p_{min} = p_V$, then, combining equations 5.3 and 5.5, it is clear that this criterion corresponds to a cavitation inception number of $\sigma_i = -C_{pmin}$. On the other hand, a departure from this criterion results in values of σ_i different from $-C_{pmin}$.

Several variations in the definition of cavitation number occur in the literature. Often the inlet tip velocity, ΩR_{T1}, is employed as the reference velocity, U, and this version will be used in this monograph unless otherwise stated. Sometimes, however, the relative velocity at the inlet tip, w_{T1}, is used as the reference velocity, U. Usually the magnitudes of w_{T1} and ΩR_{T1} do not differ greatly, and so the differences in the two cavitation numbers are small.

In the context of pumps and turbines, a number of other, surrogate cavitation parameters are frequently used in addition to some special terminology. The *NPSP* (for net positive suction pressure) is an acronym used for $(p_1^T - p_V)$, where p_1^T is the inlet total pressure given by

$$p_1^T = p_1 + \frac{1}{2}\rho v_1^2 \tag{5.6}$$

For future purposes, note from equations 5.6, 5.4, and 2.17 that

$$\left(p_1^T - p_V\right) = \frac{1}{2}\rho\Omega^2 R_{T1}^2 \left(\sigma + \phi_1^2\right) \tag{5.7}$$

Also, the *NPSE*, or net positive suction energy, is defined as $(p_1^T - p_V)/\rho$, and the *NPSH*, or net positive suction head, is $(p_1^T - p_V)/\rho g$. Furthermore, a nondimensional version of these quantities is defined in a manner similar to the specific speed as

$$S = \Omega Q^{\frac{1}{2}}/(NPSE)^{\frac{3}{4}} \tag{5.8}$$

and is called the "suction specific speed." Like the specific speed, N, the suction specific speed, is a dimensionless number, and should be computed using a consistent set of units, such as Ω in rad/s, Q in ft^3/s and $NPSE$ in ft^2/s^2. Unfortunately, it is traditional U.S. practice to use Ω in rpm, Q in gpm, and to use the $NPSH$ in ft rather than the $NPSE$. As in the case of the specific speed, one may obtain the traditional U.S. evaluation by multiplying the rational suction specific speed used in this monograph by 2734.6.

The suction specific speed is similar in concept to the cavitation number in that it represents a nondimensional version of the inlet or suction pressure. Moreover, there will be a certain critical value of the suction specific speed at which cavitation first appears. This special value is termed the inception suction specific speed, S_i. The reader should note that frequently, when a value of the "suction specific speed" is quoted for a pump, the value being given is some critical value of S that may or may not correspond to S_i. More frequently, it corresponds to S_a, the value at which the degradation in the head rise reaches a certain percentage value (see section 5.5).

The suction specific speed, S, may be obtained from the cavitation number, σ, and vice versa, by noting that, from the relations 2.17, 5.4, 5.6, and 5.8, it follows that

$$S = \left[\pi\phi_1 \left(1 - R_{H1}^2/R_{T1}^2\right)\right]^{\frac{1}{2}} / \left[\frac{1}{2}\left(\sigma + \phi_1^2\right)\right]^{\frac{3}{4}} \tag{5.9}$$

We should also make note of a third nondimensional parameter, called Thoma's cavitation factor, σ_{TH}, which is defined as

$$\sigma_{TH} = \left(p_1^T - p_V\right) / \left(p_2^T - p_1^T\right) \tag{5.10}$$

where $(p_2^T - p_1^T)$ is the total pressure rise across the pump. Clearly, this is connected to σ and to S by the relation

$$\sigma_{TH} = \frac{\sigma + \phi_1^2}{\psi} = \left(\frac{N}{S}\right)^{\frac{4}{3}} \tag{5.11}$$

Since cavitation usually occurs at the inlet to a pump, σ_{TH} is not a particularly useful parameter since $(p_2^T - p_1^T)$ is not especially relevant to the phenomenon.

5.3 Cavitation Inception

For illustrative purposes in the last section, we employed the criterion that cavitation occurs when the minimum pressure in the flow just reaches the vapor pressure, $\sigma_i = -C_{pmin}$. If this were the case, the prediction of cavitation would be a straightforward matter. Unfortunately, large departures from this criterion can occur in practice, and, in this section, we shall try to present a brief overview of the reasons for these discrepancies. There is, of course, an extensive body of literature on this subject, and we shall not attempt a comprehensive review. The reader is referred to reviews by Knapp, Daily, and Hammit (1970), Acosta and Parkin (1975), Arakeri (1979) and Brennen (1994) for more detail.

First, it is important to recognize that vapor does not necessarily form when the pressure, p, in a liquid falls below the vapor pressure, p_V. Indeed, a pure liquid can, theoretically, sustain a tension, $\Delta p = p_V - p$, of many atmospheres before nucleation, or the appearance of vapor bubbles, occurs. Such a process is termed homogeneous nucleation, and has been observed in the laboratory with some pure liquids (not water) under very clean conditions. In real engineering flows, these large tensions do not occur because vapor bubbles grow from nucleation sites either on the containing surfaces or suspended in the liquid. As in the case of a solid, the ultimate strength is determined by the weaknesses (stress concentrations) represented by the nucleation sites or "nuclei." Research has shown that suspended nuclei are more important than surface nucleation sites in determining cavitation inception. These suspended nuclei may take the form either of microbubbles or of solid particles within which, perhaps, there are microbubbles. For example, a microbubble of radius, R_N, containing only vapor, is in equilibrium when the liquid pressure

$$p = p_V - 2S/R_N \tag{5.12}$$

where S is the surface tension. It follows that such a microbubble would result in a critical tension of $2S/R_N$, and the liquid pressure would have to fall below $p_V - 2S/R_N$ before the microbubble would grow to a visible size. For example, a 10 μm bubble in water at normal temperatures leads to a tension of 0.14 bar.

It is virtually impossible to remove all the particles, microbubbles and dissolved air from any substantial body of liquid (the catch-all term "liquid quality" is used to refer to the degree of contamination). Because of this contamination, substantial differences in the inception cavitation number (and, indeed, the form of cavitation) have been observed in experiments in different water tunnels, and even in a single facility with differently processed water. The ITTC comparative tests (Lindgren and Johnsson 1966, Johnsson 1969) provided a particularly dramatic example of these differences when cavitation on the same axisymmetric headform was examined in many different water tunnels around the world. An example of the variation of σ_i in those experiments, is reproduced as figure 5.1.

Because the cavitation nuclei are crucial to an understanding of cavitation inception, it is now recognized that the liquid in any cavitation inception study must be monitored by measuring the number of nuclei present in the liquid. This information is normally presented in the form of a nuclei number distribution function, $N(R_N)$, defined such that the number of nuclei per unit total volume with radii between R_N and $R_N + dR_N$ is given by $N(R_N)dR_N$. Typical nuclei number distributions are shown in figure 5.2 where data from water tunnels and from the ocean are presented.

Most of the methods currently used for making these measurements are still in the development stage. Devices based on acoustic scattering, and on light scattering, have been explored. Other instruments, known as cavitation susceptibility meters, cause samples of the liquid to cavitate, and measure the number and size of the resulting macroscopic bubbles. Perhaps the most reliable method has been the use of holography

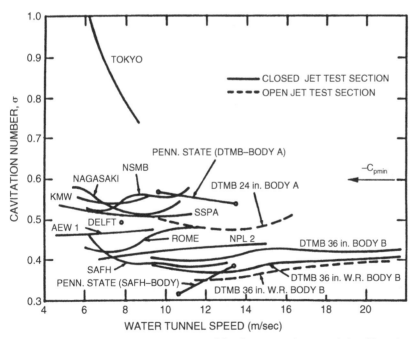

Figure 5.1. The inception numbers measured for the same axisymmetric headform in a variety of water tunnels around the world. Data collected as part of a comparative study of cavitation inception by the International Towing Tank Conference (Lindgren and Johnsson 1966, Johnsson 1969).

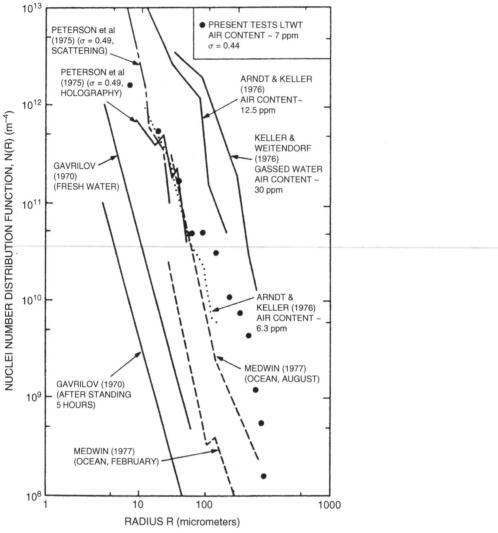

Figure 5.2. Several nuclei number distribution functions measured in water tunnels and in the ocean by various methods (adapted from Gates and Acosta 1978).

to create a magnified three-dimensional photographic image of a sample volume of liquid that can then be surveyed for nuclei. Billet (1985) has recently reviewed the current state of cavitation nuclei measurements (see also Katz *et al.* 1984).

It may be interesting to note that cavitation itself is a source of nuclei in many facilities. This is because air dissolved in the liquid will tend to come out of solution at low pressures, and contribute a partial pressure of air to the contents of any macroscopic cavitation bubble. When that bubble is convected into a region of higher pressure and the vapor condenses, this leaves a small air bubble that only redissolves very slowly, if at all. This unforeseen phenomenon caused great difficulty for the first water tunnels which were modeled directly on wind tunnels. It was discovered that, after a few minutes of operating with a cavitating body in the working section, the bubbles produced by the cavitation grew rapidly in number, and began to complete the circuit of the facility so that they appeared in the incoming flow. Soon the working section was obscured by a two-phase flow. The solution had two components. First, a water tunnel needs to be fitted with a long and deep return leg so that the water remains at high pressure for sufficient time to redissolve most of the cavitation-produced nuclei. Such a return leg is termed a "resorber." Second, most water tunnel facilities have a "deaerator" for reducing the air content of the water to 20–50% of the saturation level at atmospheric pressure. These comments serve to illustrate the fact that $N(R_N)$ in any facility can change according to the operating condition, and can be altered both by deaeration and by filtration.

Most of the data of figure 5.2 is taken from water tunnel water that has been somewhat filtered and degassed, or from the ocean which is surprisingly clean. Thus, there are few nuclei with a size greater than 100 μm. On the other hand, it is quite possible in many pump applications to have a much larger number of larger bubbles and, in extreme situations, to have to contend with a two-phase flow. Gas bubbles in the inflow could grow substantially as they pass through the low pressure regions within the pump, even though the pressure is everywhere above the vapor pressure. Such a phenomenon is called pseudo-cavitation. Though a cavitation inception number is not particularly relevant to such circumstances, attempts to measure σ_i under these circumstances would clearly yield values larger than $-C_{pmin}$.

On the other hand, if the liquid is quite clean with only very small nuclei, the tension that this liquid can sustain means that the minimum pressure has to fall well below p_V for inception to occur. Then σ_i is much smaller than $-C_{pmin}$. Thus the quality of the water and its nuclei can cause the cavitation inception number to be either larger or smaller than $-C_{pmin}$.

There are, however, at least two other factors that can affect σ_i, namely the residence time and turbulence. The residence time effect arises because the nuclei must remain at a pressure below the critical value for a sufficient length of time to grow to observable size. This requirement will depend on both the size of the pump and the speed of the flow. It will also depend on the temperature of the liquid for, as we shall see later, the rate of bubble growth may depend on the temperature of the liquid. The residence time effect requires that a finite region of the flow be below the critical pressure, and, therefore, causes σ_i to be lower than might otherwise be expected.

Up to this point we have assumed that the flow and the pressures are laminar and steady. However, most of the flows with which one must deal in turbomachinery are not only turbulent but also unsteady. Vortices occur because they are inherent in turbulence and because of both free and forced shedding of vortices. This has important consequences for cavitation inception, because the pressure in the center of a vortex may be significantly lower than the mean pressure in the flow. The measurement or calculation of $-C_{pmin}$ would elicit the value of the lowest mean pressure, while cavitation might first occur in a transient vortex whose central pressure was lower than the lowest mean pressure. Unlike the residence time factor, this would cause higher values of σ_i than would otherwise be expected. It would also cause σ_i to change with Reynolds number, Re. Note that this would be separate from the effect of Reynolds number on the minimum pressure coefficient, C_{pmin}. Note also that surface roughness can promote cavitation by creating localized low pressure perturbations in the same manner as turbulence.

5.4 Scaling of Cavitation Inception

The complexity of the issues raised in the last section helps to explain why serious questions remain as to how to scale cavitation inception. This is perhaps one of the most troublesome issues that the developer of a liquid turbomachine must face. Model tests of a ship's propeller or large turbine (to quote two common examples) may allow the designer to accurately estimate the noncavitating performance of the device. However, he will not be able to place anything like the same confidence in his ability to scale the cavitation inception data.

Consider the problem in more detail. Changing the size of the device will alter not only the residence time effect but also the Reynolds number. Furthermore, the nuclei will now be a different size relative to the impeller. Changing the speed in an attempt to maintain Reynolds number scaling may only confuse the issue by also altering the residence time. Moreover, changing the speed will also change the cavitation number, and, to recover the modeled condition, one must then change the inlet pressure which may alter the nuclei content. There is also the issue of what to do about the surface roughness in the model and in the prototype.

The other issue of scaling that arises is how to anticipate the cavitation phenomena in one liquid based on data in another. It is clearly the case that the literature contains a great deal of data on water as the fluid. Data on other liquids is quite meager. Indeed the author has not located any nuclei number distributions for a fluid other than water. Since the nuclei play such a key role, it is not surprising that our current ability to scale from one liquid to another is quite tentative.

It would not be appropriate to leave this subject without emphasizing that most of the remarks in the last two sections have focused on the inception of cavitation. Once cavitation has become established, the phenomena that occur are much less sensitive to special factors, such as the nuclei content. Hence the scaling of developed cavitation can be anticipated with much more confidence than the scaling of cavitation inception. This is not, however, of much solace to the engineer charged with avoiding cavitation completely.

5.5 Pump Performance

The performance of a pump when presented nondimensionally will take the generic form sketched in figure 5.3. As discussed earlier, the noncavitating performance will consist of the head coefficient, ψ, as a function of the flow coefficient, ϕ, where the design conditions can be identified as a particular point on the $\psi(\phi)$ curve. The noncavitating characteristic should be independent of the speed, Ω, though at lower speeds there may be some deviation due to viscous or Reynolds number effects. The *cavitating performance*, as illustrated on the right in figure 5.3, is presented as a family of curves, $\psi(\phi, \sigma)$, each for a specific flow coefficient, in a graph of the head coefficient against cavitation number, σ. Frequently, of course, both performance curves are presented dimensionally; then, for example, the NPSH is often used instead of the cavitation number as the abscissa for the cavitation performance graph.

It is valuable to identify three special cavitation numbers in the cavitation performance graph. Consider a pump operating at a particular flow rate or flow coefficient, while the inlet pressure, $NPSH$, or cavitation number is gradually reduced. As discussed in the previous chapter, the first critical cavitation number to be reached is that at which cavitation first appears; this is called the cavitation inception number, σ_i. Often the occurrence of cavitation is detected by the typical crackling sound that it makes (see section 6.5). As the pressure is further reduced, the extent (and noise) of cavitation will increase. However, it typically requires a further, substantial decrease in σ before any degradation in performance is encountered. When this occurs, the cavitation number at which it happens is often defined by a certain percentage loss in the head rise, H, or head coefficient, ψ, as shown in figure 5.3. Typically a critical cavitation number, σ_a, is defined at which the head loss is 2, 3, or 5%. Further reduction in the cavitation number will lead to major deterioration in the performance; the

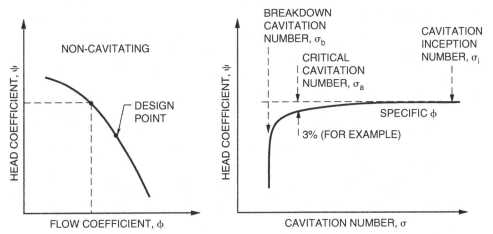

Figure 5.3. Schematic of noncavitating performance, $\psi(\phi)$, and cavitating performance, $\psi(\phi, \sigma)$, showing the three key cavitation numbers.

Table 5.1. *Inception and breakdown suction specific speeds for some*
typical pumps (from McNulty and Pearsall 1979).

PUMP TYPE	N_D	Q/Q_D	S_i	S_b	S_b/S_i
Process pump with	0.31	0.24	0.25	2.0	8.0
volute and diffuser		1.20	0.8	2.5	3.14
Double entry pump	0.96	1.00	<0.6	2.1	>3.64
with volute		1.20	0.8	2.1	2.67
Centrifugal pump with	0.55	0.75	0.6	2.41	4.02
diffuser and volute		1.00	0.8	2.67	3.34
Cooling water pump	1.35	0.50	0.65	3.40	5.24
(1/5 scale model)		0.75	0.60	3.69	6.16
		1.00	0.83	3.38	4.07
Cooling water pump	1.35	0.50	0.55	2.63	4.76
(1/8 scale model)		0.75	0.78	3.44	4.40
		1.00	0.99	4.09	4.12
		1.25	1.07	2.45	2.28
Cooling water pump	1.35	0.50	0.88	3.81	4.35
(1/12 scale model)		0.75	0.99	4.66	4.71
		1.00	0.75	3.25	4.30
		1.25	0.72	1.60	2.22
Volute pump	1.00	0.60	0.76	1.74	2.28
		1.00	0.83	2.48	2.99
		1.20	1.21	2.47	2.28

cavitation number at which this occurs is termed the breakdown cavitation number,
and is denoted by σ_b.

It is important to emphasize that these three cavitation numbers may take quite
different values, and to confuse them may lead to serious misunderstanding. For
example, the cavitation inception number, σ_i, can be an order of magnitude larger
than σ_a or σ_b. There exists, of course, a corresponding set of critical suction specific
speeds that we denote by S_i, S_a, and S_b. Some typical values of these parameters
are presented in table 5.1 which has been adapted from McNulty and Pearsall (1979).
Note the large differences between S_i and S_b.

Perhaps the most common misunderstanding concerns the recommendation of the
Hydraulic Institute that is reproduced in figure 5.4. This suggests that a pump should
be operated with a Thoma cavitation factor, σ_{TH}, in excess of the value given in the
figure for the particular specific speed of the application. The line, in fact, corresponds
to a critical suction specific speed of 3.0. Frequently, this is erroneously interpreted as
the value of S_i. In fact, it is more like S_a; operation above the line in figure 5.4 does
not imply the absence of cavitation, or of cavitation damage.

Data from McNulty and Pearsall (1979) for σ_i and σ_a in a typical pump is presented
graphically in figure 5.5 as a function of the fraction of design flow and the Reynolds

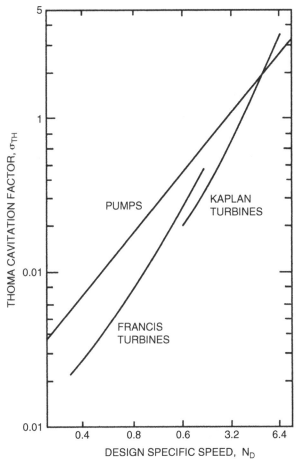

Figure 5.4. The Hydraulic Institute standards for the operation of pumps and turbines (Hydraulic Institute 1965).

number (or velocity). Note the wide scatter in the inception data, and that no clear trend with Reynolds number seems to be present.

The next section will include a qualitative description of the various forms of cavitation that can occur in a pump. Following that, the detailed development of cavitation in a pump will be described, beginning in section 5.7 with a discussion of inception.

5.6 Types of Impeller Cavitation

Since cavitation in a pump impeller can take a variety of forms (see, for example, Wood 1963), it is appropriate at this stage to attempt some description and classification of these types of cavitation. It should be borne in mind that any such classification is necessarily somewhat arbitrary, and that types of cavitation may occur that do not readily fall within the classification system. Figure 5.6 includes sketches of some of

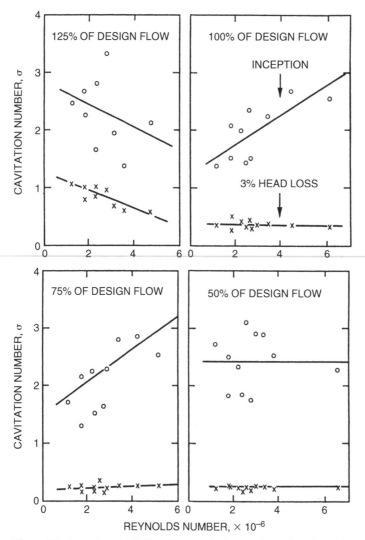

Figure 5.5. Inception and 3% head loss cavitation numbers plotted against a Reynolds number (based on w_{T1} and blade chord length) for four flow rates (from McNulty and Pearsall 1979).

the forms of cavitation that can be observed in an unshrouded axial flow impeller. As the inlet pressure is decreased, inception almost always occurs in the tip vortex generated by the corner where the leading edge meets the tip. Figure 5.7 includes a photograph of a typical cavitating tip vortex from tests of Impeller IV (the scale model of the SSME low pressure LOX turbopump shown in figure 2.12). Note that the backflow causes the flow in the vicinity of the vortex to have an upstream velocity component. Careful smoothing of the transition from the leading edge to the tip can reduce σ_i, but it will not eliminate the vortex, or vortex cavitation.

Usually the cavitation number has to be lowered quite a bit further before the next development occurs, and often this takes the form of traveling bubble cavitation on

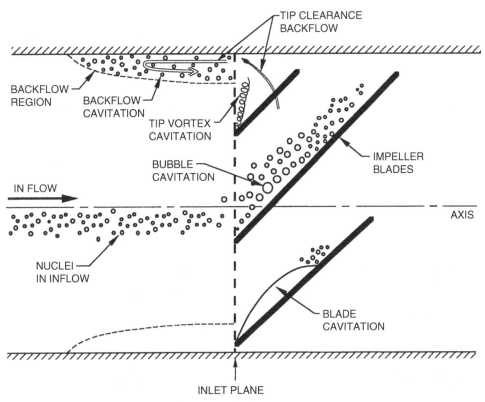

Figure 5.6. Types of cavitation in pumps.

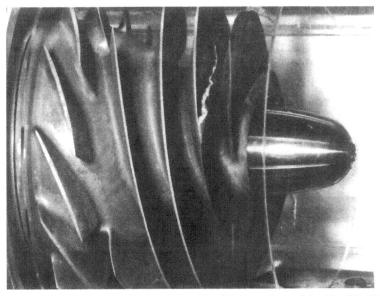

Figure 5.7. Tip vortex cavitation on Impeller IV, the scale model of the SSME low pressure LOX turbopump (see figure 2.12) at an inlet flow coefficient, ϕ_1, of 0.07 and a cavitation number, σ, of 0.42 (from Braisted 1979).

the suction surfaces of the blades. Nuclei in the inflow grow as they are convected into the regions of low pressure on the suction surfaces of the blades, and then collapse as they move into regions of higher pressure. For convenience, this will be termed "bubble cavitation." It is illustrated in figure 5.8 which shows bubble cavitation on a single hydrofoil.

With further reduction in the cavitation number, the bubbles may combine to form large attached cavities or vapor-filled wakes on the suction surfaces of the blades. In a more general context, this is known as "attached cavitation." In the context of pumps, it is often called "blade cavitation." Figure 5.9 presents an example of blade cavitation in a centrifugal pump.

Figure 5.8. Bubble cavitation on the surface of a NACA 4412 hydrofoil at zero incidence angle, a speed of 13.7 m/s and a cavitation number of 0.3. The flow is from left to right and the leading edge of the foil is just to the left of the white glare patch on the surface (Kermeen 1956).

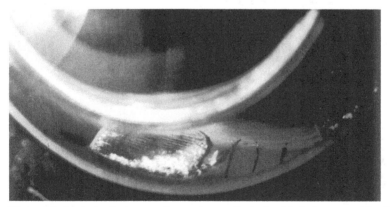

Figure 5.9. Blade cavitation on the suction surface of a blade in a centrifugal pump. The relative flow is from left to right and the cavity begins at the leading edge of the blade which is toward the left of the photograph (from Sloteman, Cooper, and Graf (1991), courtesy of Ingersoll-Dresser Pump Company).

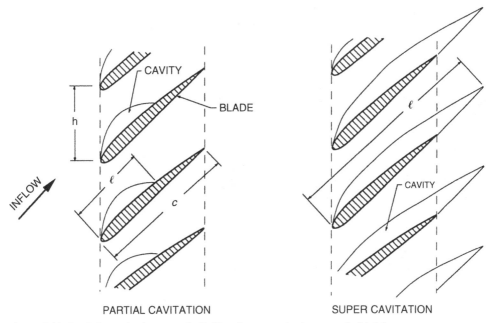

Figure 5.10. Partially cavitating cascade (left) and supercavitating cascade (right).

When blade cavities (or bubble or vortex cavities) extend to the point on the suction surface opposite the leading edge of the next blade, the increase in pressure in the blade passage tends to collapse the cavity. Consequently, the surface opposite the leading edge of the next blade is a location where cavitation damage is often encountered.

Blade cavitation that collapses on the suction surface of the blade is also referred to as "partial cavitation," in order to distinguish it from the circumstances that occur at very low cavitation numbers, when the cavity may extend into the discharge flow downstream of the trailing edge of the blade. These long cavities, which are clearly more likely to occur in lower solidity machines, are termed "supercavities." Figure 5.10 illustrates the difference between partial cavitation and supercavitation. Some pumps have even been designed to operate under supercavitating conditions (Pearsall 1963). The potential advantage is that bubble collapse will then occur downstream of the blades, and cavitation damage might thus be minimized.

Finally, it is valuable to create the catch-all term "backflow cavitation" to refer to the cavitating bubbles and vortices that occur in the annular region of backflow upstream of the inlet plane when the pump is required to operate in a loaded condition below the design flow rate (see section 4.5). The increased pressure rise across the pump under these circumstances may cause the tip clearance flow to penetrate upstream and generate a backflow that can extend many diameters upstream of the inlet plane. When the pump also cavitates, bubbles and vortices are swept up in this

Figure 5.11. As figure 5.7, but here showing typical backflow cavitation.

backflow and, to the observer, can often represent the most visible form of cavitation. Figure 5.11 includes a photograph illustrating the typical appearance of backflow cavitation upstream of the inlet plane of an inducer.

5.7 Cavitation Inception Data

In section 5.3 the important role played by cavitation nuclei in determining cavitation inception was illustrated by reference to the comparitive IFTC tests (figure 5.1). It is now clear that measurements of cavitation inception are of little value unless the nuclei population is documented. Unfortunately, this calls into question the value of most of the cavitation inception data found in the literature. And, even more important in the present context, is the fact that this includes just about all of the observations of cavitation inception in pumps. To illustrate this point, we reproduce in figure 5.12 data obtained by Keller (1974) who measured cavitation inception numbers for flows around hemispherical bodies. The water was treated in different ways so that it contained different populations of nuclei, as shown on the left in figure 5.12. As one might anticipate, the water with the higher nuclei population had a substantially larger inception cavitation number.

One of the consequences of this dependence on nuclei population is that it may cause the cavitation number at which cavitation disappears when the pressure is increased (known as the "desinent" cavitation number, σ_d) to be larger than the value at which the cavitation appeared when the pressure was decreased, namely σ_i. This phenomenon is termed "cavitation hysteresis" (Holl and Treaster 1966), and is often the result of the fact mentioned previously (section 5.3) that the cavitation itself can

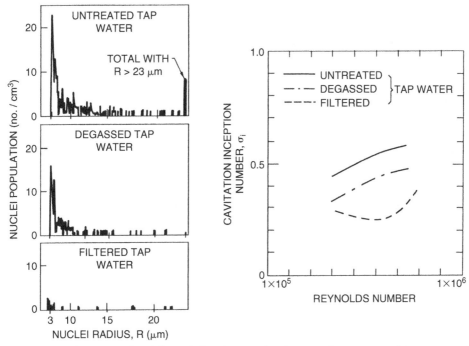

Figure 5.12. Histograms of nuclei populations in treated and untreated tap water and the corresponding cavitation inception numbers on hemispherical headforms of three different diameters (Keller 1974).

increase the nuclei population in a recirculating facility. An example of cavitation hysteresis in tests on an axial flow pump in a closed loop is given in figure 7.8.

One of the additional complications is the subjective nature of the judgment that cavitation has appeared. Visual inspection is not always possible, nor is it very objective, since the number of "events" (an event is a single bubble growth and collapse) tends to increase over a range of cavitation numbers. If, therefore, one made a judgment based on a certain critical event rate, it is inevitable that the inception cavitation number would increase with nuclei population, as in figure 5.12. Experiments have found, however, that the production of noise is a simpler and more repeatable measure of inception than visual observation. While still subject to the variations with nuclei population, it has the advantage of being quantifiable. Figure 5.13, from McNulty and Pearsall (1979), illustrates the rapid increase in the noise from a centrifugal pump when cavitation inception occurs (the data on inception in figure 5.5 and table 5.1 was obtained acoustically).

Though information on the nuclei are missing in most experiments, the total air content of the water is frequently monitored. One would suppose that the nuclei population would increase with the air content, and this is usually the case. Some data on the dependence of the critical cavitation numbers for a centrifugal pump on the total air content is included in figure 5.14. As expected, the cavitation inception number, σ_i, increases with air content. Note, however, that the breakdown cavitation

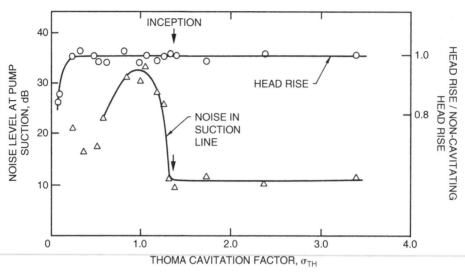

Figure 5.13. Head rise and suction line noise as a function of the Thoma cavitation factor, σ_{TH}, for a typical centrifugal pump (adapted from McNulty and Pearsall 1979).

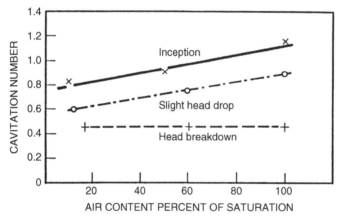

Figure 5.14. The effect of air content on the critical cavitation numbers for a centrifugal pump (Schoeneberger 1965, Pearsall 1972).

number, σ_b, is quite independent of air content, an illustration of the fact that, once it has been initiated, cavitation is much less dependent on the nuclei population.

Having begun by questioning the value of much of the cavitation inception data, we will nevertheless proceed to review some of the important trends in that data base. In doing so we might take refuge in the thought that each investigator probably applied a consistent criterion in assessing cavitation inception, and that the nuclei content in a given facility might be fairly constant (though the latter is very doubtful). Then, though the data from different investigators and facilties may be widely scattered, one would hope that the trends exhibited in a particular research project would be qualitatively significant.

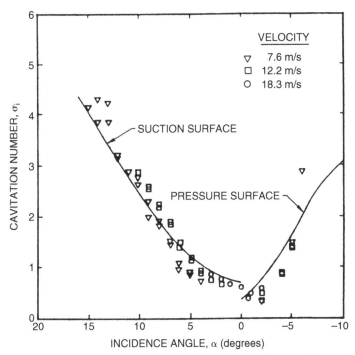

Figure 5.15. Cavitation inception characteristics of a NACA 4412 hydrofoil (Kermeen 1956).

Consider first the inception characteristics of a single hydrofoil as the angle of incidence is varied. Typical data, obtained by Kermeen (1956) for a NACA 4412 hydrofoil, is reproduced in figure 5.15. At positive angles of incidence, the regions of low pressure and cavitation inception will occur on the suction surface; at negative angles of incidence, these phenomena will shift to what is normally the pressure surface. Furthermore, as the angle of incidence is increased in either direction, the value of $-C_{pmin}$ will increase, and hence the inception cavitation number will also increase.

When such hydrofoils are used to construct a cascade, the results will also depend on the cascade solidity, s. Data on the pressure distributions around a blade in a cascade (such as that of Herrig *et al.* 1957) can be used to determine C_{pmin} as a function of blade angle, β_{b1}, solidity, s, and angle of incidence, α. Consequently, one can anticipate the variation in the inception number with these variables, assuming the first-order approximation, $\sigma_i = -C_{pmin}$. An example of such data is presented in figure 5.16; this was derived by Pearsall (1972) from the cascade data of Herrig *et al.* (1957). Note that a particular cascade will have a particular positive angle of incidence of, typically, a few degrees, at which σ_i is a minimum. The optimum angle of incidence changes with different s and β_{b1}; however, it seems to lie within a fairly narrow range between 1 and 5 degrees for a wide range of those design variables. In a pump, the incidence angle is usually small in the vicinity of the design flow rate, but will increase substantially above or below the design value. Consequently, in a pump, the cavitation inception

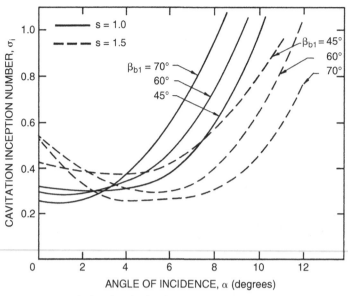

Figure 5.16. Calculated cavitation inception number, σ_i (or $-C_{pmin}$), as a function of blade angle, β_{b1}, solidity, s, and incidence angle, α, for a cascade of NACA-65-010 hydrofoils (Herrig *et al.* 1957, Pearsall 1972).

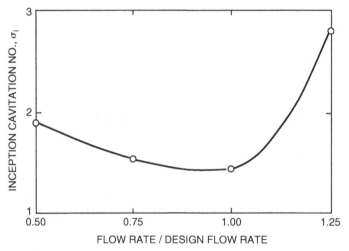

Figure 5.17. Variation in the inception number with flow rate for a typical centrifugal pump (adapted from McNulty and Pearsall 1979).

number tends to have a minimum at the design flow rate. This is illustrated in figure 5.17 which includes some data from a typical centrifugal pump, and by the data in figure 7.7 for an axial flow pump.

As we discussed in section 5.4, the scaling of cavitation phenomena with size and with speed can be an important issue. Typical data for cavitation inception on a single hydrofoil is that obtained by Holl and Wislicenus (1961); it is reproduced in

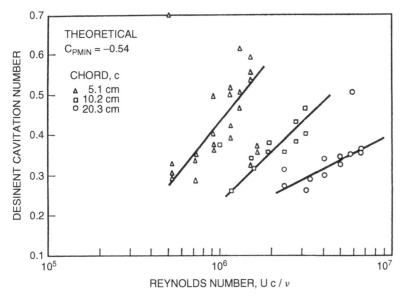

Figure 5.18. The desinent cavitation numbers for three geometrically similar Joukowski hydrofoils at zero angle of incidence as a function of Reynolds number, Uc/ν (Holl and Wislicenus 1961). Note that the theoretical $C_{pmin} = -0.54$.

figure 5.18. Data for three different sizes of 12% Joukowski hydrofoil (at zero angle of incidence) were obtained at different speeds. It was plotted against Reynolds number in the hope that this would reduce the data to a single curve. The fact that this does not occur demonstrates that there is a size or speed effect separate from that due to the Reynolds number. It seems plausible that the missing parameter is the ratio of the nuclei size to chord length; however, in the absence of information on the nuclei, such a conclusion is speculative.

To complete the list of those factors that may influence cavitation inception, it is necessary to mention the effects of surface roughness and of the turbulence level in the flow. The two effects are connected to some degree, since roughness will affect the level of turbulence. But roughness can also affect the flow by delaying boundary layer separation and therefore modifying the pressure and velocity fields in a more global manner. The reader is referred to Arndt and Ippen (1968) for details of the effects of surface roughness on cavitation inception.

Turbulence affects cavitation inception since a nucleus may find itself in the core of a vortex where the pressure level is lower than the mean. It could therefore cavitate when it might not do so under the influence of the prevailing mean pressure level. Thus turbulence may promote cavitation, but one must also allow for the fact that it may alter the global pressure field by altering the location of flow separation. These complicated viscous effects on cavitation inception were first examined in detail by Arakeri and Acosta (1974) and Gates and Acosta (1978) (see also Arakeri 1979). The

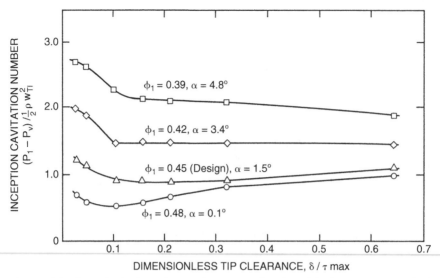

Figure 5.19. The cavitation inception number, σ_i, as a function of tip clearance, δ (τ_{max} is the maximum blade thickness), in an unshrouded axial flow pump at various flow coefficients, ϕ (adapted from Rains 1954).

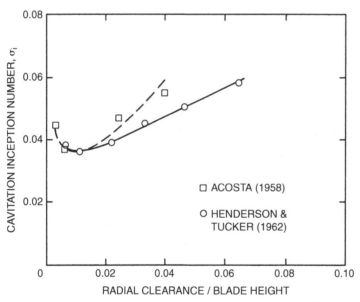

Figure 5.20. The cavitation inception number as a function of radial tip clearance in an axial inducer (Janigro and Ferrini 1973 from data of Acosta 1958 and Henderson and Tucker 1962).

implications for cavitation inception in the highly turbulent environment of most pump flows have yet to be examined in detail.

In unshrouded turbomachinery, cavitation usually begins in the vortices associated with the tip clearance flows, and so it is important to investigate how the tip clearance

will affect the inception number. In figures 5.19 and 5.20 observed cavitation inception numbers for the tip clearance flows in axial flow impellers are plotted against
nondimensional tip clearance. The typical variation with incidence angle or flow
coefficients is illustrated in figure 5.19 (Rains 1954). Since the pressure difference
between the two sides of the blade increases with incidence angle, the velocities of
the tip clearance flow must also increase, and it follows that σ_i should increase correspondingly, as is the case in figure 5.19. A second feature that is not clear in Rains'
data, but is manifest in the data of Acosta (1958) and Henderson and Tucker (1962), is
that there appears to be an optimum tip clearance of about 1% of the blade height. At
this optimum, the cavitation inception number is a minimum. This is illustrated in the
figure 5.20.

6

Bubble Dynamics, Damage and Noise

6.1 Introduction

We now turn to the characteristics of cavitation for $\sigma < \sigma_i$. To place the material in context, we begin with a discussion of bubble dynamics, so that reference can be made to some of the classic results of that analysis. This leads into a discussion of two of the deleterious effects that occur as soon as there is any cavitation, namely cavitation damage and cavitation noise. In the next chapter, we address another deleterious consequence of cavitation, namely its effect upon hydraulic performance.

6.2 Cavitation Bubble Dynamics

Two fundamental models for cavitation have been extensively used in the literature. One of these is the spherical bubble model which is most relevant to those forms of bubble cavitation in which nuclei grow to visible, macroscopic size when they encounter a region of low pressure, and collapse when they are convected into a region of higher pressure. For present purposes, we give only the briefest outline of these methods, while referring the reader to the extensive literature for more detail (see, for example, Knapp, Daily, and Hammitt 1970, Plesset and Prosperetti 1977, Brennen 1994). The second fundamental methodology is that of free streamline theory, which is most pertinent to flows consisting of attached cavities or vapor-filled wakes; a brief review of this methodology is given in chapter 7.

Virtually all of the spherical bubble models are based on some version of the Rayleigh-Plesset equation (Plesset and Prosperetti 1977) that defines the relation between the radius of a spherical bubble, $R(t)$, and the pressure, $p(t)$, far from the bubble. In an otherwise quiescent incompressible Newtonian liquid, this equation takes the form

$$\frac{p_B(t) - p(t)}{\rho_L} = R\frac{d^2R}{dt^2} + \frac{3}{2}\left(\frac{dR}{dt}\right)^2 + \frac{4\nu}{R}\frac{dR}{dt} + \frac{2\mathcal{S}}{\rho_L R} \qquad (6.1)$$

where ν, $\mathcal{S}$, and ρ_L are respectively the kinematic viscosity, surface tension, and density of the liquid. This equation (without the viscous and surface tension terms) was first derived by Rayleigh (1917) and was first applied to the problem of a traveling cavitation bubble by Plesset (1949).

The pressure far from the bubble, $p(t)$, is an input function that could be obtained from a determination of the pressure history that a nucleus would experience as it travels along a streamline. The pressure, $p_B(t)$, is the pressure inside the bubble. It is often assumed that the bubble contains both vapor and noncondensable gas, so that

$$p_B(t) = p_V(T_B) + \frac{3 m_G K_G T_B}{4\pi R^3} = p_V(T_\infty) - \rho_L \Theta + \frac{3 m_G K_G T_B}{4\pi R^3} \qquad (6.2)$$

where T_B is the temperature inside the bubble, $p_V(T_B)$ is the vapor pressure, m_G is the mass of gas in the bubble, and K_G is the gas constant. However, it is convenient to use the ambient liquid temperature far from the bubble, T_∞, to evaluate p_V. When this · is done, it is necessary to introduce the term, Θ, into equation 6.1 in order to correct the difference between $p_V(T_B)$ and $p_V(T_\infty)$. It is this term, Θ, that is the origin of the thermal effect in cavitation. Using the Clausius-Clapeyron relation,

$$\Theta \cong \frac{\rho_V \mathcal{L}}{\rho_L T_\infty} (T_\infty - T_B(t)) \qquad (6.3)$$

where ρ_V is the vapor density and $\mathcal{L}$ is the latent heat.

Note that in using equation 6.2 for $p_B(t)$, we have introduced the additional unknown function, $T_B(t)$, into the Rayleigh-Plesset equation 6.1. In order to determine this function, it is necessary to construct and solve a heat diffusion equation, and an equation for the balance of heat in the bubble. Approximate solutions to these equations can be written in the following simple form. If the heat conducted into the bubble is equated to the rate of use of latent heat at the interface, then

$$\left(\frac{\partial T}{\partial r}\right)_{r=R} = \frac{\rho_V \mathcal{L}}{k_L} \frac{dR}{dt} \qquad (6.4)$$

where $(\partial T/\partial r)_{r=R}$ is the temperature gradient in the liquid at the interface and k_L is the thermal conductivity of the liquid. Moreover, an approximate solution to the thermal diffusion equation in the liquid is

$$\left(\frac{\partial T}{\partial r}\right)_{r=R} = \frac{(T_\infty - T_B(t))}{(\alpha_L t)^{\frac{1}{2}}} \qquad (6.5)$$

where α_L is the thermal diffusivity of the liquid ($\alpha_L = k_L/\rho_L c_{PL}$ where c_{PL} is the specific heat of the liquid) and t is the time from the beginning of bubble growth or collapse. Using equations 6.4 and 6.5 in equation 6.3, the thermal term can be approximated as

$$\Theta = \Sigma(T_\infty) t^{\frac{1}{2}} \frac{dR}{dt} \qquad (6.6)$$

where

$$\Sigma(T_\infty) = \frac{\rho_V^2 \mathcal{L}^2}{\rho_L^2 c_{PL} T_\infty \alpha_L^{\frac{1}{2}}} \tag{6.7}$$

In section 7.7, we shall utilize these relations to evaluate the thermal suppression effects in cavitating pumps.

For present purposes, it is useful to illustrate some of the characteristic features of solutions to the Rayleigh-Plesset equation in the absence of thermal effects ($\Theta = 0$ and $T_B(t) = T_\infty$). A typical solution of $R(t)$ for a nucleus convected through a low pressure region is shown in figure 6.1. Note that the response of the bubble is quite nonlinear; the growth phase is entirely different in character from the collapse phase. The growth is steady and controlled; it rapidly reaches an asymptotic growth rate in which the dominant terms of the Rayleigh-Plesset equation are the pressure difference, $p_V - p$, and the second term on the right-hand side so that

$$\frac{dR}{dt} \Rightarrow \left[\frac{2(p_V - p)}{3\rho_L} \right]^{\frac{1}{2}} \tag{6.8}$$

Note that this requires the local pressure to be less than the vapor pressure. For traveling bubble cavitation, the typical tension $(p_V - p)$ will be given nondimensionally by

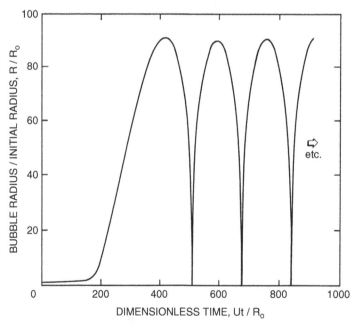

Figure 6.1. Typical solution, $R(t)$, of the Rayleigh-Plesset equation for a spherical bubble originating from a nucleus of radius, R_0. The nucleus enters a low pressure region at a dimensionless time of 0 and is convected back to the original pressure at a dimensionless time of 500. The low pressure region is sinusoidal and symmetric about a dimensionless time of 250.

$(-C_{pmin} - \sigma)$ (see equations 5.2 and 5.4) so the typical growth rate is given by

$$\frac{dR}{dt} \propto (-C_{pmin} - \sigma)^{\frac{1}{2}} U \tag{6.9}$$

While this growth rate may appear, superficially, to represent a relatively gentle process, it should be recognized that it corresponds to a volume that is increasing like t^3. Cavitation growth is therefore an explosive process to be contrasted with the kind of boiling growth that occurs in a kettle on the stove in which dR/dt typically behaves like $t^{-\frac{1}{2}}$. The latter is an example of the kind of thermally inhibited growth discussed in section 7.7.

It follows that we can estimate the typical maximum size of a cavitation bubble, R_M, given the above growth rate and the time available for growth. Numerical calculations using the full Rayleigh-Plesset equation show that the appropriate time for growth is the time for which the bubble experiences a pressure below the vapor pressure. In traveling bubble cavitation we may estimate this by knowing the shape of the pressure distribution near the minimum pressure point. We shall represent this shape by

$$C_p = C_{pmin} + C_{p*}(s/D)^2 \tag{6.10}$$

where s is a coordinate measured along the surface, D is the typical dimension of the body or flow, and C_{p*} is some known constant of order one. Then, the time available for growth, t_G, is given approximately by

$$t_G \approx \frac{2D \left(\sigma - C_{pmin}\right)^{\frac{1}{2}}}{C_{p*}^{\frac{1}{2}} U \left(1 + C_{pmin}\right)^{\frac{1}{2}}} \tag{6.11}$$

and therefore

$$\frac{R_M}{D} \approx \frac{2 \left(-\sigma - C_{pmin}\right)}{C_{p*}^{\frac{1}{2}} \left(1 + C_{pmin}\right)^{\frac{1}{2}}} \tag{6.12}$$

Note that this is independent of the size of the original nucleus.

One other feature of the growth process is important to mention. It transpires that because of the stabilizing influence of the surface tension term, a particular tension, $(p_V - p)$, will cause only bubbles larger than a certain critical size to grow explosively (Blake 1949). This means that, for a given cavitation number, only nuclei larger than a certain critical size will achieve the growth rate necessary to become macroscopic cavitation bubbles. A decrease in the cavitation number will activate smaller nuclei, thus increasing the volume of cavitation. This phenomenon is illustrated in figure 6.2 which shows the maximum size of a cavitation bubble, R_M, as a function of the size of the original nucleus and the cavitation number for a typical flow around an

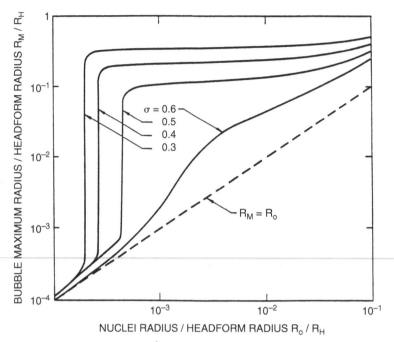

Figure 6.2. The maximum size to which a cavitation bubble grows (according to the Rayleigh-Plesset equation), R_M, as function of the original nuclei size, R_0, and the cavitation number, σ, in the flow around an axisymmetric headform of radius, R_H, with $\mathcal{S}/\rho_L R_H U^2 = 0.000036$ (from Ceccio and Brennen 1991).

axisymmetric headform. The vertical parts of the curves on the left of the figure represent the values of the critical nuclei size, R_C, that are, incidentally, given simply by the expression

$$R_C \approx \kappa \mathcal{S}/\rho_L U^2 \left(-\sigma - C_{pmin}\right) \qquad (6.13)$$

where the factor κ is roughly unity (Ceccio and Brennen 1991). Note also from figure 6.2 that all the unstable nuclei grow to roughly the same size as anticipated earlier.

Turning now to the collapse, it is readily seen from figure 6.1 that cavitation bubble collapse is a catastrophic phenomenon in which the bubble, still assumed spherical, reaches a size very much smaller than the original nucleus. Very high accelerations and pressures are generated when the bubble becomes very small. However, if the bubble contains any noncondensable gas at all, this will cause a rebound as shown in figure 6.1. Theoretically, the spherical bubble will undergo many cycles of collapse and rebound. In practice, a collapsing bubble becomes unstable to nonspherical disturbances, and essentially shatters into many smaller bubbles in the first collapse and rebound. The resulting cloud of smaller bubbles rapidly disperses. Whatever the deviations from the spherical shape, the fact remains that the collapse is a violent process that produces noise and the potential for material damage to nearby surfaces. We proceed to examine both of these consequences in the two sections which follow.

6.3 Cavitation Damage

Perhaps the most ubiquitous problem caused by cavitation is the material damage that cavitation bubbles can cause when they collapse in the vicinity of a solid surface. Consequently, this aspect of cavitation has been intensively studied for many years (see, for example, ASTM 1967, Knapp, Daily, and Hammitt 1970, Thiruvengadam 1967, 1974). The problem is complex because it involves the details of a complicated unsteady flow combined with the reaction of the particular material of which the solid surface is made.

As we have seen in the previous section, cavitation bubble collapse is a violent process that generates highly localized, large amplitude disturbances and shocks in the fluid at the point of collapse. When this collapse occurs close to a solid surface, these intense disturbances generate highly localized and transient surface stresses. Repetition of this loading due a multitude of bubble collapses can cause local surface fatigue failure, and the detachment of pieces of material. This is the generally accepted explanation for cavitation damage. It is consistent with the appearance of cavitation damage in most circumstances. Unlike the erosion due to solid particles in the flow, for which the surface appears to be smoothly worn with scratches due to larger particles, cavitation damage has the crystalline and jagged appearance of fatigue failure. To illustrate this, a photograph of localized cavitation damage on the blade of a mixed flow pump, fabricated from an aluminium-based alloy, is included as figure 6.3. More extensive damage is illustrated in figure 6.4 which shows the blades at discharge from a Francis turbine; here the cavitation damage has penetrated the blades. Cavitation damage can also occur in much larger scale flows. As an example, figure 6.5 shows

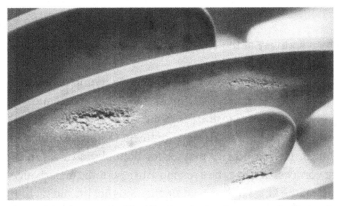

Figure 6.3. Photograph of localized cavitation damage on the blade of a mixed flow pump impeller made from an aluminium-based alloy.

Figure 6.4. Cavitation damage on the blades at the discharge from a Francis turbine.

Figure 6.5. Cavitation damage to the concrete wall of the 15.2 m diameter Arizona spillway at the Hoover Dam. The hole is 35 m long, 9 m wide, and 13.7 m deep. Reproduced from Warnock (1945).

cavitation damage suffered by a spillway at the Hoover dam (Warnock 1945, Falvey 1990).

In hydraulic devices such as pump impellers or propellers, cavitation damage is often observed to occur in quite localized areas of the surface. This is frequently the result of the periodic and coherent collapse of a cloud of cavitation bubbles. Such is the case in magnetostrictive cavitation testing equipment (Knapp, Daily, and Hammitt 1970). In many pumps, the periodicity may occur naturally as a result of regular shedding of cavitating vortices, or it may be a response to a periodic disturbance imposed on the flow. Examples of the kinds of imposed fluctuations are the interaction between a row of rotor vanes and a row of stator vanes, or the interaction between a ship's propeller and the nonuniform wake behind the ship. In almost all such cases, the

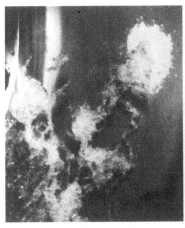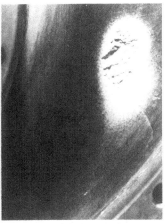

Figure 6.6. Axial views from the inlet of the cavitation and cavitation damage on the hub or base plate of a centrifugal pump impeller. The two photographs are of the same area, the left one showing the typical cavitation pattern during flow and the right one the typical cavitation damage. Parts of the blades can be seen in the upper left and lower right corners; relative to these blades the flow proceeds from the lower left to the upper right. The leading edge of the blade is just outside the field of view on the upper left. Reproduced from Soyama, Kato, and Oba (1992) with permission of the authors.

coherent collapse of the cloud can cause much more intense noise and more potential for damage than in a similar nonfluctuating flow. Consequently, the damage is most severe on the solid surface close to the location of cloud collapse. An example of this phenomenon is included in figure 6.6 taken from Soyama, Kato, and Oba (1992). In this instance, clouds of cavitation are being shed from the leading edge of a centrifugal pump blade, and are collapsing in a specific location as suggested by the pattern of cavitation in the left-hand photograph. This leads to the localized damage shown in the right-hand photograph.

Currently, several research efforts are focussed on the dynamics of cavitation clouds. These studies suggest that the coherent collapse can be more violent than that of individual bubbles, but the basic explanation for the increase in the noise and damage potential is not clear.

6.4 Mechanism of Cavitation Damage

The intense disturbances that are caused by cavitation bubble collapse can have two separate origins. The first is related to the fact that a collapsing bubble may be unstable in terms of its shape. When the collapse occurs near a solid surface, Naude and Ellis (1961) and Benjamin and Ellis (1966) observed that the developing spherical asymmetry takes the form of a rapidly accelerating jet of fluid, entering the bubble from the side furthest from the wall (see figure 6.7). Plesset and Chapman (1971) carried out numerical calculations of this "reentrant jet," and found good agreement with the experimental observations of Lauterborn and Bolle (1975). Since then, other

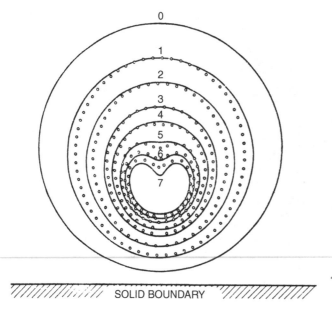

Figure 6.7. The collapse of a cavitation bubble close to a solid boundary. The theoretical shapes of Plesset and Chapman (1971) (solid lines) are compared with the experimental observations of Lauterborn and Bolle (1975) (points) (adapted from Plesset and Prosperetti 1977).

analytical methods have explored the parametric variations in the flow. These methods are reviewed by Blake and Gibson (1987). The "microjet" achieves very high speeds, so that its impact on the other side of the bubble generates a shock wave, and a highly localized shock loading of the surface of the nearby wall.

Parenthetically, we might remark that this is also the principle on which the depth charge works. The initial explosion creates little damage, but does produce a very large bubble which, when it collapses, generates a reentrant jet directed toward any nearby solid surface. When this surface is a submarine, the collapse of the bubble can cause great damage to that vessel. It may also be of interest to note that a bubble, collapsing close to a very flexible or free surface, develops a jet on the side closest to this boundary, and, therefore, traveling in the opposite direction. Some researchers have explored the possibility of minimizing cavitation damage by using surface coatings with a flexibility designed to minimize the microjet formation.

The second intense disturbance occurs when the remnant cloud of bubbles, that remains after the microjet disruption, collapses to its minimum gas/vapor volume, and generates a second shock wave that impinges on the nearby solid surface. The generation of a shock wave during the rebound phase of bubble motion was first demonstrated by the calculations of Hickling and Plesset (1964). More recently, Shima *et al.* (1981) have made interesting observations of the spherical shock wave using Schlieren photography, and Fujikawa and Akamatsu (1980) have used photoelastic solids to examine the stress waves developed in the solid. Though they only observed stress waves resulting from the remnant cloud collapse and not from the microjet, Kimoto (1987) has subsequently shown that both the microjet and the remnant cloud create stress waves in the solid. His measurements indicate that the surface loading resulting from the remnant cloud is about two or three times that due to the microjet.

Until very recently, virtually all of these detailed observations of collapsing cavitation bubbles had been made in a quiescent fluid. However, several recent observations have raised doubts regarding the relevance of these results for most flowing systems. Ceccio and Brennen (1991) have made detailed observations of the collapse of cavitating bubbles in flows around bodies, and have observed that typical cavitation bubbles are distorted and often broken up by the shear in the boundary layer or by the turbulence before the collapse takes place. Furthermore, Chahine (personal communication) has performed calculations similar to those of Plesset and Chapman, but with the addition of rotation due to shear, and has found that the microjet is substantially modified and reduced by the flow.

The other important facet of the cavitation damage phenomenon is the reaction of the material of the solid boundary to the repetitive shock (or "water hammer") loading. Various measures of the resistance of particular materials to cavitation damage have been proposed (see, for example, Thiruvengadam 1967). These are largely heuristic and empirical, and will not be reviewed here. The reader is referred to Knapp, Daily, and Hammitt (1970) for a detailed account of the relative resistance of different materials to cavitation damage. Most of these comparisons are based, not on tests in flowing systems, but on results obtained when material samples are vibrated at high frequency (about $20\ kHz$) in a bath of quiescent liquid. The samples are weighed

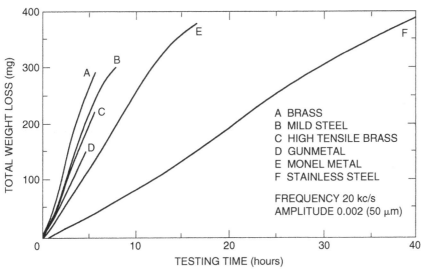

Figure 6.8. Examples of cavitation damage weight loss as a function of time. Data from vibratory tests with different materials (Hobbs, Laird and Brunton 1967).

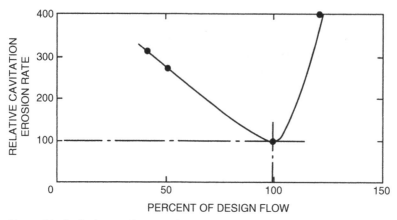

Figure 6.9. Cavitation erosion rates in a centrifugal pump as a function of the flow rate relative to the design flow rate (Pearsall 1978 from Grist 1974).

at regular intervals to determine the loss of material, and the results are presented in the form typified by figure 6.8. Note that the relative erosion rates, according to this data, can be approximately correlated with the structural strength of the material. Furthermore, the erosion rate is not necessarily constant with time. This may be due to the differences in the response of a collapsing bubble to a smooth surface as opposed to a surface already roughened by damage. Finally, note that the weight loss in many materials only begins after a certain incubation time.

The data on erosion rates in pumps is very limited because of the length of time necessary to make such measurements. The data that does exist (Mansell 1974) demonstrates that the rate of erosion is a strong function of the operating point as given by the cavitation number and the flow coefficient. The influence of the latter is illustrated in figure 6.9. This curve essentially mirrors those of figures 5.16 and 5.17. At off-design conditions, the increased angle of incidence leads to increased cavitation and, therefore, increased weight loss.

6.5 Cavitation Noise

The violence of cavitation bubble collapse also produces noise. In many practical circumstances, the noise is important not only because of the vibration that it may cause, but also because it advertizes the presence of cavitation and, therefore, the likelihood of cavitation damage. Indeed, the magnitude of cavitation noise is often used as a crude measure of the rate of cavitation erosion. For example, Lush and Angell (1984) have shown that, in a given flow at a given cavitation number, the rate of weight loss due to cavitation damage correlates with the noise as the velocity of the flow is changed.

Prior to any discussion of cavitation noise, it is useful to identify the natural frequency with which individual bubbles will oscillate a quiescent liquid. This natural frequency can be obtained from the Rayleigh-Plesset equation 6.1 by substituting an expression for $R(t)$ that consists of a constant, R_E, plus a small sinusoidal perturbation of amplitude, $\tilde{R}$, at a general frequency, ω. Steady state oscillations like this would only be maintained by an applied pressure, $p(t)$, consisting of a constant, $\bar{p}$, plus a sinusoidal perturbation of amplitude, $\tilde{p}$, and frequency, ω. Obtaining the relation between the linear perturbations, $\tilde{R}$ and $\tilde{p}$, from the Rayleigh-Plesset equation, it is found that the ratio, $\tilde{R}/\tilde{p}$, has a maximum at a resonant frequency, ω_P, given by

$$\omega_P = \left[\frac{3(\bar{p} - p_V)}{\rho_L R_E^2} + \frac{4S}{\rho_L R_E^3} - \frac{8\nu^2}{R_E^4} \right]^{\frac{1}{2}} \tag{6.14}$$

The results of this calculation for bubbles in water at $300^\circ K$ are presented in figure 6.10 for various mean pressure levels, $\bar{p}$. Note that the bubbles below about $0.02 \ \mu m$ are supercritically damped, and have no resonant frequency. Typical cavitation nuclei of size $10 \rightarrow 100 \ \mu m$ have resonant frequencies in the range $10 \rightarrow 100 \ kHz$. Even though the nuclei are excited in a highly nonlinear way by the cavitation, one might expect that the spectrum of the noise that this process produces would have a broad maximum at the peak frequency corresponding to the size of the most numerous nuclei participating in the cavitation. Typically, this would correspond to the radius of the critical nucleus given by the expression 6.13. For example, if the critical nuclei size were of the order of 10–$100 \ \mu m$, then, according to figure 6.10, one might expect

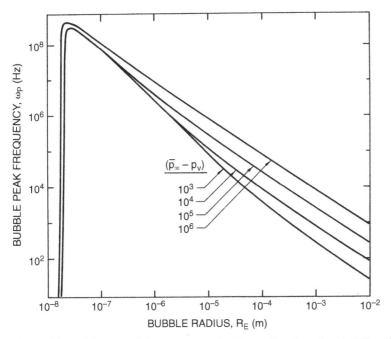

Figure 6.10. Bubble natural frequency, ω_P, in Hz as a function of the bubble radius and the difference between the equilibrium pressure and the vapor pressure (in $kg/m \ sec^2$) for water at $300^\circ K$.

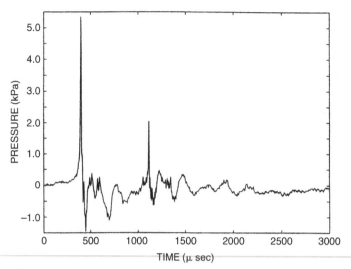

Figure 6.11. Typical acoustic signal from a single collapsing bubble (from Ceccio and Brennen 1991).

to see cavitation noise frequencies of the order of $10-100\,kHz$. This is, indeed, the typical range of frequencies produced by cavitation.

Fitzpatrick and Strasberg (1956) were the first to make extensive use of the Rayleigh-Plesset equation to predict the noise from individual collapsing bubbles and the spectra that such a process would produce. More recently, Ceccio and Brennen (1991) have recorded the noise from individual cavitation bubbles in a flow. A typical acoustic signal is reproduced in figure 6.11. The large positive pulse at about $450\;\mu s$ corresponds to the first collapse of the bubble. Since the radiated acoustic pressure, p_A, in this context is related to the second derivative of the volume of the bubble, $V(t)$, by

$$p_A = \frac{\rho_L}{4\pi\ell}\frac{d^2V}{dt^2} \tag{6.15}$$

(where ℓ is the distance of the measurement from the center of the bubble), the pulse corresponds to the very large and positive values of d^2V/dt^2 that occur when the bubble is close to its minimum size in the middle of the collapse. The first pulse is followed in figure 6.11 by some facility-dependent oscillations, and by a second pulse at about $1100\;\mu s$. This corresponds to the second collapse; no further collapses were observed in these particular experiments.

A good measure of the magnitude of the collapse pulse in figure 6.11 is the acoustic impulse, I, defined as the area under the curve or

$$I = \int_{t_1}^{t_2} p_A dt \tag{6.16}$$

where t_1 and t_2 are the times before and after the pulse when $p_A = 0$. The acoustic impulses for cavitation on two axisymmetric headforms (ITTC and Schiebe

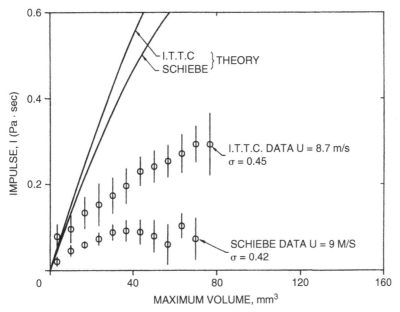

Figure 6.12. The acoustic impulse, I, produced by the collapse of a single cavitation bubble. Data is shown for two axisymmetric bodies (the ITTC and Schiebe headforms) as a function of the maximum volume prior to collapse. Also shown are the equivalent results from solutions of the Rayleigh-Plesset equation (from Ceccio and Brennen 1991).

headforms) are compared in figure 6.12 with impulses predicted from integration of the Rayleigh-Plesset equation. Since these theoretical calculations assume that the bubble remains spherical, the discrepancy between the theory and the experiments is not too surprising. Indeed, the optimistic interpretation of figure 6.12 is that the theory can provide an order of magnitude estimate of the noise produced by a single bubble. This could then be combined with the nuclei number density distribution to obtain a measure of the amplitude of the noise (Brennen 1994).

The typical single bubble noise shown in figure 6.11 leads to the spectrum shown in figure 6.13. If the cavitation events are randomly distributed in time, this would also correspond to the overall cavitation noise spectrum. It displays a characteristic frequency content in the range of $1 \rightarrow 50 \ kHz$ (the rapid decline at about $80 \ kHz$ represents the limit of the hydrophone used to make these measurements). Typical measurements of the noise produced by cavitation in an axial flow pump are illustrated in figure 6.14, and exhibit the same features demonstrated in figure 6.13. The signal in figure 6.14 also clearly contains some shaft or blade passage frequencies that occur in the absence of cavitation, but may be amplified or attenuated by cavitation. Figure 6.15 contains data obtained for cavitation noise in a centrifugal pump. Note that the noise at a frequency of $40 \ kHz$ shows a sharp increase with the onset of cavitation; on the other hand, the noise at the shaft and blade passage frequencies show only minor changes with cavitation number. The decrease in the $40 \ kHz$ cavitation

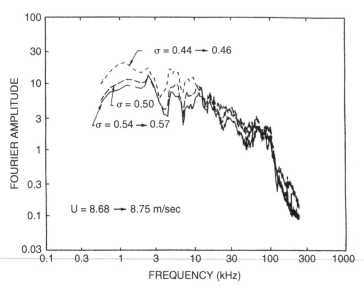

Figure 6.13. Typical spectra of noise from bubble cavitation for various cavitation numbers as indicated (Ceccio and Brennen 1991).

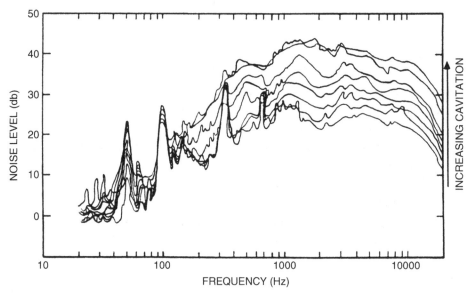

Figure 6.14. Typical spectra showing the increase in noise with increasing cavitation in an axial flow pump (Lee 1966).

noise as breakdown is approached is also a common feature in cavitation noise measurements.

The level of the sound produced by a cavitating flow is the result of two factors, namely the impulse, I, produced by each event (equation 6.16) and the event rate or number of events per second, $\dot{N}_E$. Therefore, the sound pressure level, p_S,

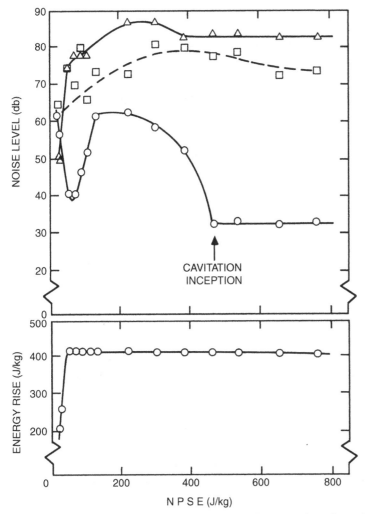

Figure 6.15. The relation between the cavitation performance, the noise and vibration produced at three frequency levels in a centrifugal pump, namely the shaft frequency ($\square$), the blade passage frequency ($\triangle$), and $40 \, kHz$ ($\circ$) (Pearsall 1966–67).

will be

$$p_S = I \dot{N}_E \tag{6.17}$$

Here, we will briefly discuss the scaling of the two components, I, and $\dot{N}_E$, and thus the scaling of the cavitation noise, p_S. We emphasize that the following equations omit some factors of proportionality necessary for quantitative calculations.

Both the experimental observations and the calculations based on the Rayleigh-Plesset equation, show that the nondimensional impulse from a single cavitation event, defined by

$$I^* = 4\pi I \ell / \rho U D^2 \tag{6.18}$$

(where U and D are the reference velocity and length in the flow), is strongly correlated with the maximum volume of the cavitation bubble (maximum equivalent volumetric radius = R_M), and appears virtually independent of the other flow parameters. In dimensionless terms,

$$I^* \approx R_M^2 / D^2 \tag{6.19}$$

It follows that

$$I \approx \rho U R_M^2 / \ell \tag{6.20}$$

The evaluation of the impulse from a single event is then completed by some estimate of the maximum bubble size, R_M. For example, we earlier estimated R_M for traveling bubble cavitation (equation 6.12), and found it to be independent of U for a given cavitation number. In that case I is linear in U.

Modeling the event rate, $\dot{N}_E$, can be considerably more complicated than might, at first sight, be visualized. If all the nuclei flowing through a certain known streamtube (say with a cross-sectional area, A_N, in the upstream reference flow), were to cavitate similarly then, clearly, the result would be

$$\dot{N}_E = N A_N U \tag{6.21}$$

where N is the nuclei concentration (number/unit volume). Then the sound pressure level resulting from substituting the expressions 6.21, 6.20, and 6.12 into equation 6.17, is

$$p_S \approx \rho U^2 \left(-\sigma - C_{pmin}\right)^2 A_N N D^2 / \ell \tag{6.22}$$

where we have omitted some of the constants of order unity. For the simple circumstances outlined, equation 6.22 yields a sound pressure level that scales with U^2 and with D^4 (because $A_N \propto D^2$). This scaling with velocity does correspond to that often observed (for example, Blake, Wolpert, and Geib 1977, Arakeri and Shangumanathan 1985) in simple traveling bubble flows. There are, however, a number of complicating factors. First, as we have discussed earlier in section 6.2, only those nuclei larger than a certain critical size, R_C, will actually grow to become cavitation bubbles, and, since R_C is a function of both σ and the velocity U, this means that N will be a function of R_C and U. Since R_C decreases as U increases, the power law dependence of p_S on velocity will then be U^m where m is greater than 2.

Different scaling laws will apply when the cavitation is generated by turbulent fluctuations, such as in a turbulent jet (see, for example, Ooi 1985, Franklin and

McMillan 1984). Then the typical tension and the typical duration of the tension experienced by a nucleus, as it moves along an approximately Lagrangian path in the turbulent flow, are very much more difficult to estimate. Consequently, estimates of the sound pressure due to cavitation in turbulent flows, and the scaling of that sound with velocity, are more poorly understood.

7

Cavitation and Pump Performance

7.1 Introduction

In this chapter we turn our attention to another of the deleterious consequences of cavitation, namely its effect upon the steady state hydraulic performance of a pump. In the next section we present several examples of the effect of cavitation on conventional pumps. This is followed by a discussion of the performance and design of cavitating inducers which are devices added to conventional pumps for the purpose of improving the cavitation performance. Subsequent sections deal with the analytical methods available for the evaluation of cavitation performance and with the thermodynamic effects of the phase change process on that performance.

7.2 Typical Pump Performance Data

A typical non-cavitating performance characteristic for a centrifugal pump is shown in figure 7.1 for the Impeller X/Volute A combination (Chamieh 1983) described in section 2.8. The design flow coefficient for this pump is $\phi_2 = 0.092$ but we note that it performs reasonably well down to about 30% of this design flow. This flexibility is characteristic of centrifugal pumps. Data is presented for three different shaft speeds, namely 600, 800, and 1200 rpm; since these agree closely we can conclude that there is no perceptible effect of Reynolds number for this range of speeds. The effect of a different volute is also illustrated by the data for Volute B which is a circular volute of circumferentially uniform area. In theory this circular volute is not well matched to the impeller discharge flow and the result is that, over most of the range of flow coefficient, the hydraulic performance is inferior to that with Volute A. However, Volute B is superior at high flow coefficients. This suggests that the flow in Volute A may be more pathological than one would like at these high flow coefficients (see sections 4.4 and 4.6). It further serves to emphasize the importance of a volute (or diffuser) and the need for an understanding of the flow in a volute at both design and off-design conditions.

Typical cavitation performance characteristics for a centrifugal pump are presented in figure 7.2 for the Impeller X/Volute A combination. The breakdown cavitation numbers in the range $\sigma = 0.1 \rightarrow 0.4$ are consistent with the data in table 5.1. Note

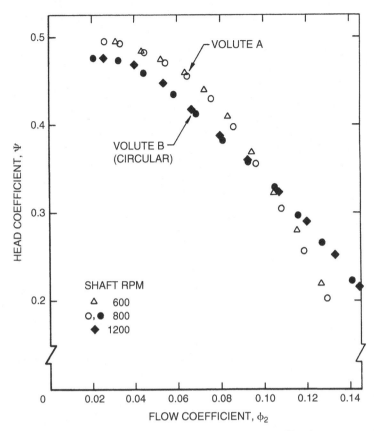

Figure 7.1. Typical non-cavitating performance for a centrifugal pump, namely Impeller X (see section 2.8) with Volute A and a circular volute of uniform cross-section (from Chamieh 1983).

that the cavitation head loss occurs more gradually at high flow coefficients than at low values. This is a common feature of the cavitation performance of many pumps, both centrifugal and axial.

Now consider some examples of axial and mixed flow pumps. Typical non-cavitating performance characteristics are shown in figure 7.3 for a Peerless axial flow pump. This unshrouded pump has a design flow coefficient $\phi_2 = 0.171$. The maximum efficiency at this design point is about 85%. Axial flow pumps are more susceptible to flow separation and stall than centrifugal pumps and could therefore be considered less versatile. The depression in the head curve of figure 7.3 in the range $\phi_2 = 0.08 \rightarrow 0.12$ is indicative of flow separation and this region of the head/flow curve can therefore be quite sensitive to the details of the blade profile since small surface irregularities can often have a substantial effect on separation. This is illustrated by the data of figure 7.4 which presents the non-cavitating characteristics for four similar axial flow pumps with slightly different blade profiles. The kinks in the curves are more marked in this case and differ significantly from one profile to another. Also note that there are small regions of positive slope in the head characteristics. This often

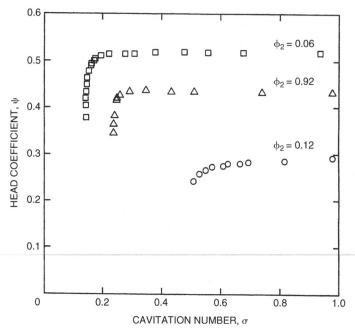

Figure 7.2. Cavitation performance for the Impeller X/Volute A combination (from Franz *et al.* 1989, 1990). The flow separation rings of figure 10.17 have been installed so the non-cavitating performance is slightly better than in figure 7.1.

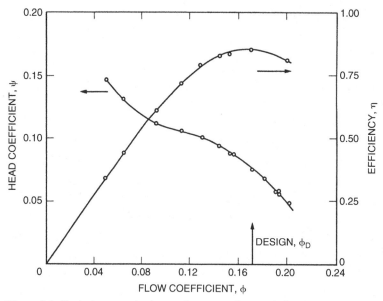

Figure 7.3. Typical non-cavitating performance characteristics for a 20.3 *cm* diameter, 3-bladed axial flow pump with a hub-tip ratio, R_H/R_T, of 0.45 running at about 1500 *rpm*. At the blade tip the chord is 7.3 *cm*, the solidity is 0.344, and the blade angle, β_{bT}, is 11.9°. Adapted from Guinard *et al.* (1953).

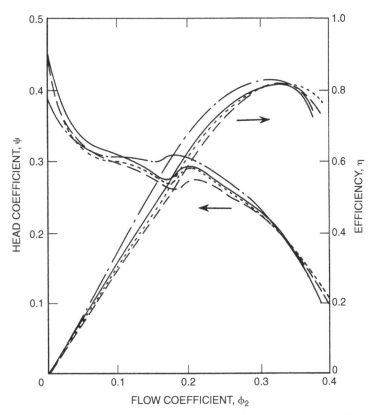

Figure 7.4. Typical non-cavitating performance characteristics for a 4-bladed axial flow pump with tip blade angle, β_{bT}, of about 18°, a hub-tip ratio, R_H/R_T, of 0.483, a solidity of 0.68 and four different blade profiles (yielding the set of four performance curves). Adapted from Oshima and Kawaguchi (1963).

leads to instability and to fluctuating pressures and flow rates through the excitation of the surge and stall mechanisms discussed in the following chapters. Sometimes the region of positive slope in the head characteristic can be even more marked as in the example presented in figure 7.5 in which the stall occurs at about 80% of the design flow. As a final example of non-cavitating performance we include in figure 7.6, the effect of the blade angle in an axial flow pump; note that angles of the order of 20° to 30° seem to be optimal for many purposes.

The cavitation characteristics for some of the above axial flow pumps are presented in figures 7.7 through 7.10. The data of Guinard *et al.* (1953) provides a particularly well-documented example of the effect of cavitation on an axial flow pump. Note first from figure 7.7 that the cavitation inception number is smallest at the design flow and increases as ϕ is decreased; the decrease at very low ϕ does not, however, have an obvious explanation. Since Guinard *et al.* (1953) noticed the hysteretic effect described in section 5.7 we present figure 7.8 as an example of that phenomenon.

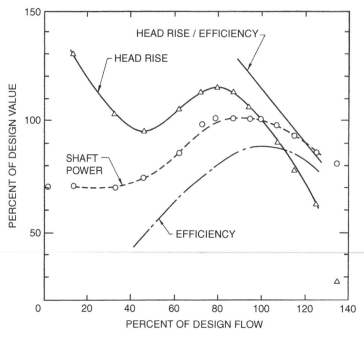

Figure 7.5. Characteristics of a mixed flow pump (Myles 1966).

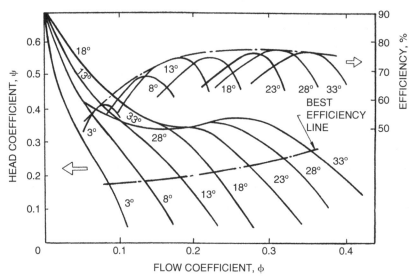

Figure 7.6. Head and efficiency characteristics for an axial flow pump with different tip blade angles, β_{bT} (from Peck 1966).

The cavitation data of figure 7.7 also help to illustrate several other characteristic phenomena. Note the significant increase in the head just prior to the decrease associated with breakdown. In the case of the pump tested by Guinard *et al.*, this effect occurs at low flow coefficients. However, other pumps exhibit this phenomenon at

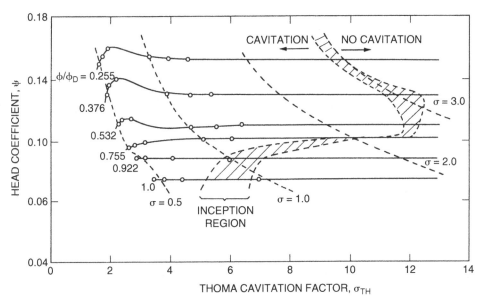

Figure 7.7. Cavitation performance characteristics of the axial flow pump of figure 7.3. Adapted from Guinard *et al.* (1953).

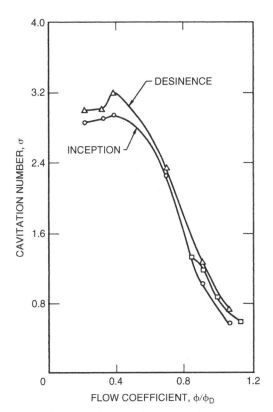

Figure 7.8. Inception and desinent cavitation numbers (based on w_{T1}) as a function of ϕ/ϕ_D for the axial flow pump of figures 7.3 and 7.7. Adapted from Guinard *et al.* (1953).

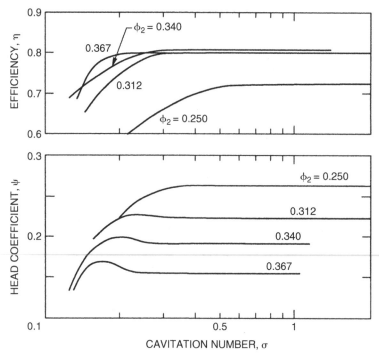

Figure 7.9. Effect of cavitation on the head coefficient and efficiency of one of the axial flow pumps of figure 7.4. The cavitation number is based on w_{T1}. Adapted from Oshima and Kawaguchi (1963).

higher flows and not at low flows as illustrated by the data of Oshima and Kawaguchi (1963) presented in figure 7.9. The effect is probably caused by an improved flow geometry due to a modest amount of cavitation.

The cavitation data of figure 7.7 also illustrates the fact that breakdown at low flow coefficients occurs at higher cavitation numbers and is usually more abrupt than at higher flow coefficients. It is accompanied by a decrease in efficiency as illustrated by figure 7.9. Finally we include figure 7.10 which shows that the effect of blade profile changes on the head breakdown cavitation number is quite small.

7.3 Inducer Designs

Axial flow inducers are intended to improve the cavitation performance of centrifugal or mixed flow pumps by increasing the inlet pressure to the pump to a level at which it can operate without excessive loss of performance due to cavitation. Typically they consist of an axial flow stage placed just upstream of the inlet to the main impeller. They are designed to operate at small incidence angles and to have thin blades so that the perturbation to the flow is small in order to minimize the production of cavitation and its deleterious effect upon the flow. The objective is to raise the pressure very gradually to the desired level. The typical advantage gained by the addition of an inducer is illustrated in figure 7.11 taken from Janigro and Ferrini (1973). This

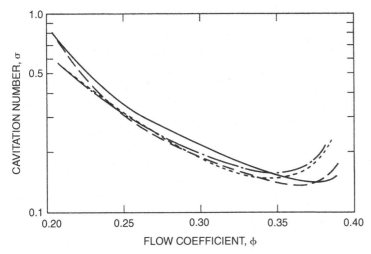

Figure 7.10. The critical cavitation number (based on w_{T1} and 0.5% head loss) for the axial flow pumps of figure 7.4. Adapted from Oshima and Kawaguchi (1963).

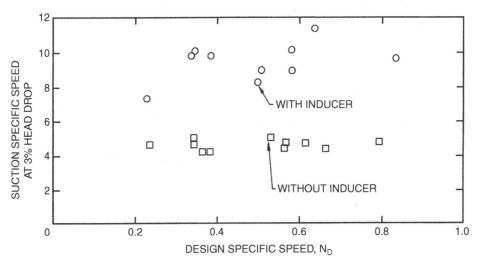

Figure 7.11. Comparison of the suction specific speed at 3% head drop for process pumps with and without inducer (from Janigro and Ferrini 1973).

compares the cavitation performance of a class of process pumps with and without an inducer.

Various types of inducer design are documented in figure 7.12 and table 7.1, both taken from Jakobsen (1971). Data on the low pressure LOX pump in the Space Shuttle Main Engine (SSME) has been added to table 7.1. Most inducers of recent design seem to be of types (a) or (b). They are unshrouded, with a swept leading edge and often with a forward cant to the blades as in the case of the low pressure LOX pump in the SSME (figure 2.12). This blade cant has the effect of causing the leading edge to be located at a single axial plane counteracting the effect of the sweep given to the leading edge.

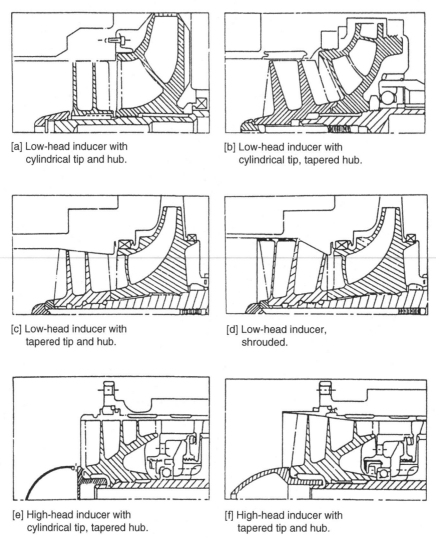

[a] Low-head inducer with
    cylindrical tip and hub.

[b] Low-head inducer with
    cylindrical tip, tapered hub.

[c] Low-head inducer with
    tapered tip and hub.

[d] Low-head inducer,
    shrouded.

[e] High-head inducer with
    cylindrical tip, tapered hub.

[f] High-head inducer with
    tapered tip and hub.

Figure 7.12. Various geometries of cavitating inducers (from Jakobsen 1971).

They are also designed to function at an incidence angle of a few degrees. The reason
that the design incidence angle is not zero is that under these conditions cavitation
could form on either the pressure or suction surfaces or it could oscillate between the
two. It is preferable to use a few degrees of incidence to eliminate this uncertainty
and ensure suction surface cavitation.

7.4 Inducer Performance

Typical inducer performance characteristics are presented in figures 7.13 through
7.16. The non-cavitating performance of simple 9° helical inducers (see figure 2.12)
is presented in figure 7.13. The data for the 5.1 *cm* and 7.6 *cm* diameter models
appear to coincide indicating very little Reynolds number effect. Furthermore, the

Table 7.1. *Typical rocket engine inducer geometry and performance (from Jakobsen 1971 and other sources). Key: (a) Main + Partial or Main/Tandem (b) Radial (RAD), Swept Backwards (SWB), or Swept Forward (SWF)*

Rocket: Fluid:	THOR LOX	J-2 LOX	X-8 LOX	X-8 LOX	J-2 LH2	J-2 LH2	SSME LOX
No. of Blades (a)	4	3	3	2	4+4	4+4	4/12
R_{H1}/R_{T1}	0.31	0.20	0.23	~0.19	0.42	0.38	0.29
R_{T2}/R_{T1}	1.0	1.0	~0.9	~0.8	1.0	~0.9	1.0
R_{H2}/R_{H1}	1.0	~2	~1.5	1.5	~2	~2	2.6
Leading Edge (b)	RAD	SWB	SWB	SWF	SWB	SWB	SWB
β_{bT1} (deg.)	14.15	9.75	9.8	5.0	7.9	7.35	7.3
ϕ_{1D}	0.116	0.109	0.106	0.05	0.094	0.074	0.076
ψ_{1D}	0.075	0.11	0.10	0.063	0.21	0.20	0.366
N_{1D}	4.21	3.06	3.25	3.15	1.75	1.61	0.68
α_{T1} (deg.)	7.5	3.5	3.7	2.1	2.5	3.1	4.3
σ_D	0.028	0.021	0.025	0.007	0.011	0.011	
S_D	10.4	12.5	11.4	21.2	15.8	16.2	

non-cavitating performance is the same whether the leading edge is swept or straight. Also included in the figure are the results of the lossless performance prediction of equation 4.6. The agreement with the experiments is about as good as one could expect. It is most satisfactory close to the zero incidence flow coefficient of about $0.09 \rightarrow 0.10$ where one would expect the viscous losses to be a minimum. The comparison also suggests that the losses increase as one either increases or decreases the flow from that zero incidence value.

The cavitation performance of the 7.58 *cm* model of the 9° helical inducer is presented in figure 7.14. These curves for different flow coefficients exhibit the typical pattern of a more gradual head loss at the higher flow coefficients. Notice that the breakdown cavitation number is smaller for non-zero incidence (for example, $\phi = 0.052$) than it is for zero incidence ($\phi = 0.095$). One would expect the breakdown cavitation number to be a minimum at zero incidence. The fact that the data do not reflect this expectation may be due to the complications at low flow coefficients caused by backflow and the prerotation which backflow induces (see section 4.5).

Another example of inducer performance is presented in figures 7.15 and 7.16, in this case for the SSME low pressure LOX pump model designated Impeller IV (see figure 2.12). In figure 7.15, non-cavitating performance characteristics are shown for two models with diameters of 7.58 *cm* and 10.2 *cm*. The difference in the two characteristics is not related to the size as much as it is to the fact that the 7.58 *cm* model was tested with a set of diffuser (stator) vanes in the axial flow annulus just downstream of the impeller discharge whereas the 10.2 *cm* model was tested without such a diffuser. Note the substantial effect that this has upon the performance. Below

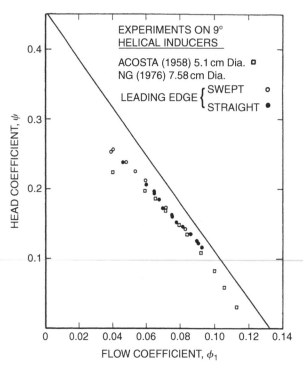

Figure 7.13. Non-cavitating performance of 9° helical inducers of two different sizes and with and without swept leading edges (the 7.58 *cm* inducers are Impellers III and V). Also shown is the theoretical performance prediction in the absence of losses (from Ng and Brennen 1978).

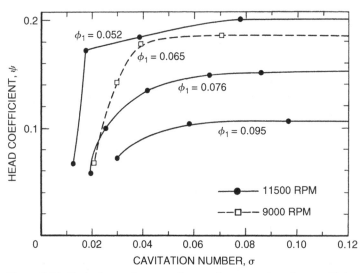

Figure 7.14. Cavitation performance for Impeller V at various flow coefficients and rotating speeds (from Ng and Brennen 1978).

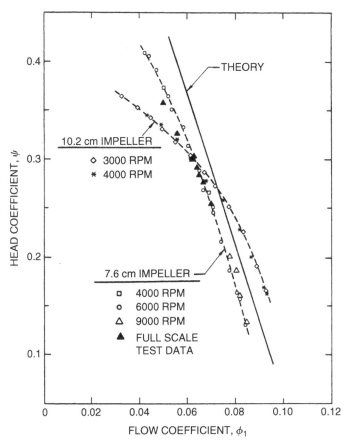

Figure 7.15. Non-cavitating performance of Impeller IV (7.58 *cm*, with stator) and Impeller VI (10.2 *cm*, without stator) at various rotational speeds. Also shown are full scale test data from Rocketdyne and a theoretical performance prediction (solid line) (from Ng and Brennen 1978).

the design flow ($\phi_1 \approx 0.076$) the stator vanes considerably improve the diffusion process. However, above the design flow, the negative angle of incidence of the flow encountering the stator vanes appears to cause substantial loss and results in degradation of the performance. Some full scale test data (with diffuser) obtained by Rocketdyne is included in figure 7.15 and shows quite satisfactory agreement with the 7.58 *cm* model tests. The results of the theoretical performance given by equation 4.6 are also shown and the comparison between the lossless theory and the experimental data is similar to that of figure 7.13.

The cavitation performance of Impeller IV in water is shown in figure 7.16 along with some data from full scale tests. Note that the head tends to be somewhat erratic at the lower cavitation numbers. Such behavior is typical of most axial flow inducer data and is probably due to hydraulic losses caused by unsteadiness in the flow (see section 8.7).

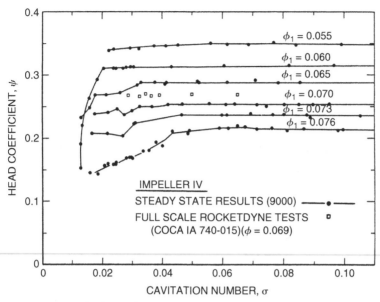

Figure 7.16. Cavitation performance of Impeller IV at 9000 rpm and various flow coefficients. Also shown are full scale test data from Rocketdyne (from Ng and Brennen 1978).

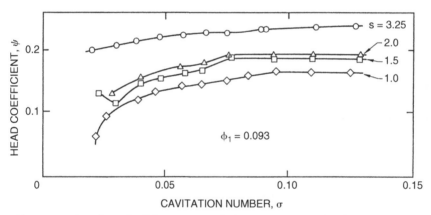

Figure 7.17. The effect of solidity on the cavitation performance of a 9° helical inducer (from Acosta 1958).

7.5 Effects of Inducer Geometry

In this section we comment on several geometric factors for which the data suggests optimum values. Clearly, the solidity, s, needs to be as small as possible and yet large enough to achieve the desired discharge flow angle. Data on the effect of the solidity on the performance of a 3-bladed, 9° helical inducer has been obtained by Acosta (1958) and on a 4-bladed, $8\frac{1}{2}°$ helical inducer by Henderson and Tucker (1962). This data is shown in figures 7.17 and 7.18. The effect on the non-cavitating performance (extreme

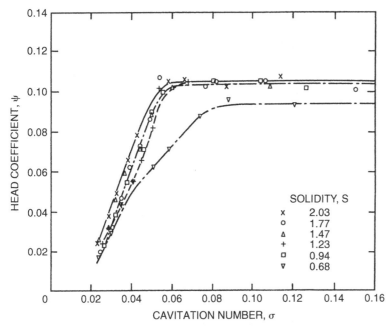

Figure 7.18. The effect of solidity on the cavitation performance of a cavitating inducer (Janigro and Ferrini 1973 from Henderson and Tucker 1962).

right of the figures) seems greater for Acosta's inducer than for that of Henderson and Tucker. The latter data suggests that, as expected, the non-cavitating performance is little affected unless the solidity is less than unity. Both sets of data suggest that the cavitating performance is affected more than the non-cavitating performance by changes in the solidity when the latter is less than about unity. Consequently, this data suggests an optimum value of s of about 1.5.

The same two studies also investigated the effect of the tip clearance and the data of Henderson and Tucker (1962) is reproduced in figure 7.19. As was the case with the solidity, the non-cavitating performance is less sensitive to changes in the tip clearance than is the cavitation performance. Note from figure 7.19 that the non-cavitating performance is relatively insensitive to the clearance unless the latter is increased above 2% of the chord when the performance begins to decline more rapidly. The cavitating performance shows a similar dependence though the fractional changes in the performance are larger. Note that the performance near the knee of the curve indicates an optimum clearance of about 1% of the chord which is in general qualitative agreement with the effect of tip clearance on cavitation inception discussed earlier (see figure 5.20).

Moore and Meng (1970a,b) have made a study of the effect of the leading edge geometry on inducer performance and their results are depicted graphically in figures 7.20 and 7.21. Note that the leading edge geometry has a significant effect on the non-cavitating performance and on the breakdown cavitation number. Simply stated, the

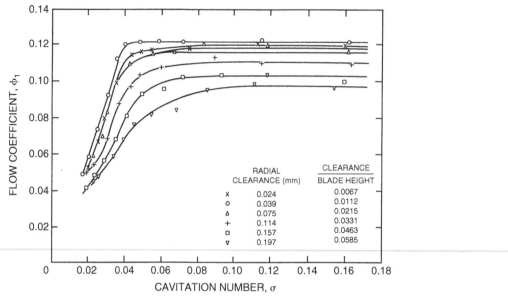

Figure 7.19. The effect of tip clearance on the cavitation performance of a cavitating inducer (from Henderson and Tucker 1962 as given by Janigro and Ferrini 1973).

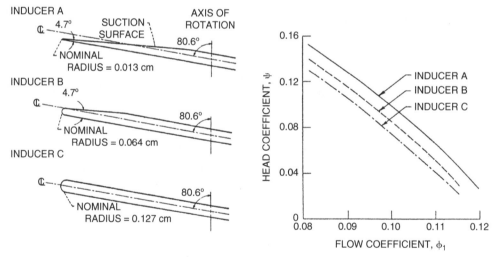

Figure 7.20. Non-cavitating performance of three 9.4° helical inducers with different leading edges as shown. Tests performed with liquid hydrogen (from Moore and Meng 1970b).

sharper the leading edge the better the hydraulic performance under both cavitating and non-cavitating conditions. There is, however, a trade-off to be made here for very thin leading edges may flutter. This phenomenon is discussed in section 8.12. Incidentally,

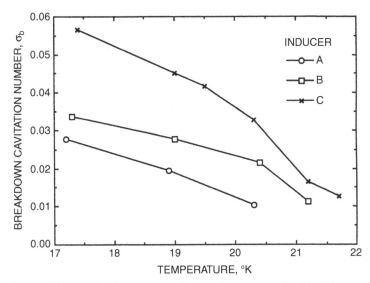

Figure 7.21. The breakdown cavitation numbers, σ_b (defined in this case by a 30% head drop) as a function of temperature for three shapes of leading edge (see figure 7.20) on 9.4° helical inducers operating in liquid hydrogen (from Moore and Meng 1970a,b).

figure 7.21 also demonstrates the thermal effect on cavitation performance which is discussed in section 7.7.

7.6 Analyses of Cavitation in Pumps

In this and the sections which follow we shall try to give a brief overview of the various kinds of models which have been developed for the analysis of developed cavitation in a pump. Clearly different types of cavitation require different analytical models. We begin in this section with the various attempts which have been made to model traveling bubble cavitation in a pump and to extract from such a model information regarding the damage potential, noise or performance decrement caused by that cavitation. In a later section we shall outline the methods developed for attached blade cavitation. As for the other types of cavitation which are typically associated with the secondary flows (for example, tip vortex cavitation, backflow cavitation) there is little that can be added to what has already been described in the last chapter. Much remains to be understood concerning secondary flow cavitation, perhaps because some of the more important effects involve highly unsteady and transient cavitation.

To return to a general discussion of traveling bubble cavitation, it is clear that, given the pressure and velocity distribution along a particular streamline in a reference frame fixed in the impeller, one can input that information into the Rayleigh-Plesset equation 6.1 as discussed in section 6.2. The equation can then be integrated to find the size of the bubble at each point along its trajectory (see examples in section 6.2). Such programs are equally applicable to two-phase flows or to two-component gas/liquid flows.

Since the first applications of the Rayleigh-Plesset equation to traveling bubble cavitation by Plesset (1949) and Parkin (1952) there have been many such investigations, most of which are reviewed by Holl (1969). A notable example is the work of Johnson and Hsieh (1966) who included the motion of the bubble relative to the liquid and demonstrated the possibility of some screening effects because of the motion of the bubbles across streamlines due to centripetal forces. While most of the literature discusses single bubble solutions of this type for flows around simple headforms the same programs can readily be used for the flow around a pump blade provided the pressure distributions on streamlines are known either from an analytic or numerical solution or from experimental measurements of the flow in the absence of cavitation. Such investigations would allow one to examine both the location and intensity of bubble collapse in order to learn more about the potential for cavitation damage.

These methods, however, have some serious limitations. First, the Rayleigh-Plesset equation is only valid for spherical bubbles and collapsing bubbles lose their spherical symmetry as discussed in section 6.4. Consequently any investigation of damage requires considerations beyond those of the Rayleigh-Plesset equation. Secondly, the analysis described above assumes that the concentration of bubbles is sufficiently small so that bubbles do not interact and are not sufficiently numerous to change the flow field from that for non-cavitating flow. This means that they are of little value in predicting the effect of cavitation on pump performance since such an effect implies interactions between the bubbles and the flow field.

It follows that to model the performance loss due to traveling bubble cavitation one must use a two-phase or two-component flow model which implicitly includes interaction between the bubbles and the liquid flow field. One of the first models of this kind was investigated by Cooper (1967) and there have been a number of similar investigations for two-component flows in pumps, for example, that by Rohatgi (1978). While these investigations are useful, they are subject to serious limitations. In particular, they assume that the two-phase mixture is in thermodynamic equilibrium. Such is certainly not the case in cavitation flows where an expression like the Rayleigh-Plesset equation is needed to describe the dynamics of disequilibrium. Nevertheless the models of Cooper and others have value as the first coherent attempts to evaluate the effects of traveling bubble cavitation on pump performance.

Rather than the assumption of thermodynamic equilibrium, two-phase bubble flow models need to be developed in which the bubble dynamics are included through appropriate use of the Rayleigh-Plesset equation. In recent years a number of investigators have employed such models to investigate the dynamics and acoustics of clouds of cavitation bubbles in which the bubbles and the flow interact (see, for example, Chahine 1982, d'Agostino and Brennen 1983, 1989, d'Agostino et al. 1988,

Biesheuval and van Wijngaarden 1984, Omta 1987). Among other things these investigations demonstrate that a cloud of bubbles has a set of natural frequencies of its own, separate from (but related to) the bubble natural frequency and that the bubble and flow interaction effects become important when the order of magnitude of the parameter $\alpha A^2/R^2$ exceeds unity where α is the void fraction and A and R are the dimensions of the cloud and bubbles respectively. These more appropriate models for traveling bubble cavitation have not, as yet, been used to investigate cavitation effects in pumps.

Several other concepts should be mentioned before we leave the subject of bubbly cavitation in pumps. One such concept which has not received the attention it deserves was put forward by Jakobsen (1964). He attempted to merge the free streamline models (which are discussed later in section 7.8) with his observations that attached cavities on the suction surfaces of impeller blades tend to break up into bubbly mixtures near the closure or reattachment point of the attached cavity. Jakobsen suggested that condensation shocks occur in this bubbly mixture and constitute a mechanism for head breakdown.

There are also a number of results and ideas that emerge from studies of the pumping of bubbly gas/liquid mixtures. One of the most important of these is found in the measurements of bubble size made by Murakami and Minemura (1977, 1978). It transpires that, in most practical pumping situations, the turbulence and shear at inlet tend to break up all the gas bubbles larger than a certain size during entry to the blade passages. The ratio of the force tending to cause fission to the surface tension, S, which tends to resist fission will be a Weber number and Murakami and Minemura (1977, 1978) suggest that the ratio of the diameter of the largest bubbles to survive the inlet shear, $2R_M$, to the blade spacing, h_1, will be a function of a Weber number, $We = \rho \Omega^2 R_{T1}^2 h_1/S$. Figure 7.22 presents some data on $2R_M/h_1$ taken by Murakami and Minemura for both centrifugal and axial flow pumps.

The size of the bubbles in the blade passages is important because it is the migration and coalesence of these bubbles that appears to cause degradation in the performance. Since the velocity of the relative motion between the bubbles and the liquid is proportional to the bubble size raised to some power which depends on the Reynolds number regime, it follows that the larger the bubbles the more likely it is that large voids will form within the blade passage due to migration of the bubbles toward regions of lower pressure (Furuya 1985, Furuya and Maekawa 1985). As Patel and Runstadler (1978) observed during experiments on centrifugal pumps and rotating passages, regions of low pressure occur not only on the suction sides of the blades but also under the shroud of a centrifugal pump. These large voids can cause substantial changes in the deviation angle of the flow leaving the impeller and hence alter the pump performance in a significant way. This mechanism of head degradation is probably significant not only for gas/liquid flows but also for cavitating flows. In gas/liquid flows the higher the velocity the greater the degree of bubble fission at inlet and the smaller the bubbles. But the force acting on the bubbles is also greater for the higher velocity flows and so the net result is not obvious. One can only conclude that both processes, inlet fission and blade passage migration, may be important and deserve further study along the lines begun by Murakami and Minemura.

At the beginning of this section we discussed the application of the Rayleigh-Plesset equation to study the behavior of individual cavitating bubbles. One area in which such an analysis has been useful is in evaluating the differences in the cavitation occurring in different liquids and in the same liquid at different temperatures. These issues will be addressed in the next section. In the subsequent section we turn our attention to the free streamline methods which have been developed to model the flows which occur when large attached cavities or gas-filled voids occur on the blades of a turbomachine.

7.7 Thermal Effect on Pump Performance

Changes in the temperature of the liquid being pumped will clearly affect the vapor pressure, p_V, and therefore the NPSH or cavitation number. This effect has, of course, already been incorporated in the analysis or presentation of the performance by using the difference between the inlet pressure and the vapor pressure rather than the absolute value of the inlet pressure as a flow parameter. But there is another effect of the liquid temperature which is not so obvious and requires some discussion and analysis. It is illustrated by figure 7.23 which includes cavitation performance data for a centrifugal pump (Arndt 1981 from Chivers 1969) operating with water at different inlet temperatures. Note that the cavitation breakdown decreases substantially with increasing temperature. Somewhat counter-intuitively the performance actually improves as the

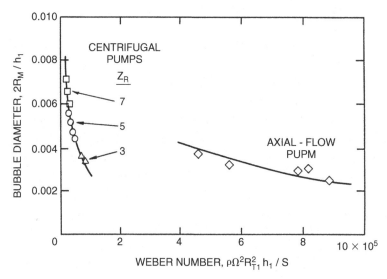

Figure 7.22. The bubble diameters observed in the blade passages of centrifugal and axial flow pumps as a function of Weber number (adapted from Murakami and Minemura 1978).

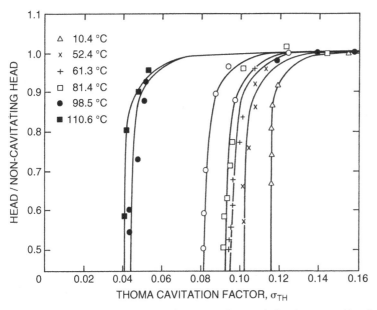

Figure 7.23. Typical cavitation performance characteristics for a centrifugal pump pumping water at various temperatures as indicated (Arndt 1981 from Chivers 1969).

temperature gets greater! The variation of the breakdown cavitation number, σ_b, with the inlet temperature in Chivers' (1969) experiments is shown in figure 7.24 which includes data for two different speeds and shows a consistent decrease in σ_b with increasing temperature. The data for the two speeds deviate somewhat at the lowest temperatures. To illustrate that the thermal effect occurs in other liquids and in other kinds of pumps, we include in figure 7.25 data reported by Gross (1973) from tests of the Saturn J-2 liquid oxygen inducer pump. This shows the same pattern manifest in figure 7.24. Other data of this kind has been obtained by Stepanoff (1961), Spraker (1965), and Salemann (1959) for a variety of other liquids.

The explanation for this effect is most readily given by making reference to traveling bubble cavitation though it can be extended to other forms of cavitation. However, for simplicity, consider a single bubble (or nucleus) which begins to grow when it enters a region of low pressure. Liquid on the surface of the bubble will vaporize to provide the increase in volume of vapor filling the bubble. Consider, now, what happens at two different temperatures, one "high" and one "low." At "low" temperatures the density of the saturated vapor is low and, therefore, the mass rate of evaporation of liquid needed is small. Consequently, the rate at which heat is needed as latent heat to effect this vaporization is low. Since the heat will be conducted from the bulk of the liquid and since the rate of heat transfer is small, this means that the amount by which the temperature of the interface falls below the bulk liquid temperature is also small. Consequently the vapor pressure in the cavity only falls slightly below the value of the vapor pressure at the bulk liquid temperature. Therefore, the driving force behind the

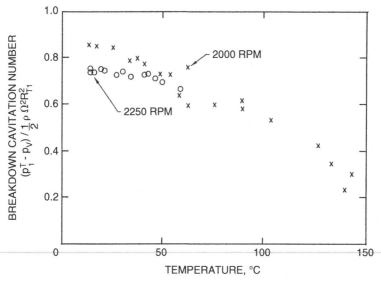

Figure 7.24. Thermodynamic effect on cavitation breakdown for a commercial centrifugal pump (data from Chivers 1969).

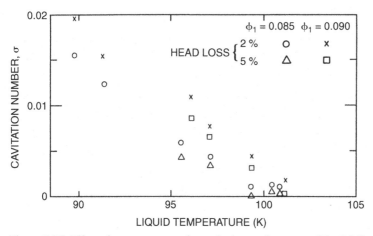

Figure 7.25. Effect of temperature on the cavitation performance of the J-2 liquid oxygen inducer pump (adapted from Gross 1973).

bubble growth, namely the difference between the internal pressure (vapor pressure) and the pressure far from the bubble, is not much influenced by thermal effects.

Now, consider the same phenomenon occuring at the "high" temperature. Since the vapor density can be many orders of magnitude larger than at the "low" temperature, the mass rate of evaporation for the same volume growth rate is much larger. Thus the heat which must be conducted to the interface is much larger which means that a substantial thermal boundary layer builds up in the liquid at the interface. This causes the temperature in the bubble to fall well below that of the bulk liquid and this, in turn, means that the vapor pressure within the bubble is much lower than otherwise might be expected. Consequently, the driving force behind the bubble growth is reduced. This reduction in the rate of bubble growth due to thermal effects is the origin of the thermal effect on the cavitation performance in pumps. Since the cavitation head loss is primarily due to disruption of the flow by volumes of vapor growing and collapsing within the pump, any reduction in the rate of bubble growth will lessen the disruption and result in improved performance.

This thermal effect can be extended to attached or blade cavities with only minor changes in the details. At the downstream end of a blade cavity, vapor is entrained by the flow at a certain volume rate which will depend on the flow velocity and other geometric parameters. At higher temperatures this implies a larger rate of entrainment of mass of vapor due to the larger vapor density. Since vaporization to balance this entrainment is occurring over the surface of the cavity, this implies a larger temperature difference at the higher temperature. And this implies a lower vapor pressure in the cavity than might otherwise be expected and hence a larger "effective" cavitation number. Consequently the cavitation performance is improved at the higher temperature.

Both empirical and theoretical arguments have been put forward in attempts to quantify these thermal effects. We shall begin with the theoretical arguments put forward by Ruggeri and Moore (1969) and by Brennen (1973). These explicitly apply to bubble cavitation and proceed as follows.

At the beginning of bubble growth, the rate of growth rapidly approaches the value given by equation 6.8 and the important $(dR/dt)^2$ term in the Rayleigh-Plesset equation 6.1 is roughly constant. On the other hand, the thermal term, Θ, which is initially zero, will grow like $t^{\frac{1}{2}}$ according to equation 6.6. Consequently there will be a critical time, t_C, at which the thermal term, Θ, will approach the magnitude of $(p_B(T_\infty) - p)/\rho_L$ and begin to reduce the rate of growth. Using the expression 6.6, this critical time is given by

$$t_C \approx (p_B - p)/\rho_L \Sigma^2 \tag{7.1}$$

For $t \ll t_C$, the dominant terms in equation 6.1 are $(p_B - p)/\rho_L$ and $(dR/dt)^2$ and the bubble growth rate is as given by equation 6.8. For $t \gg t_C$, the dominant terms in equation 6.1 become $(p_B - p)/\rho_L$ and Θ so that, using equations 6.4 and 6.5, the bubble growth rate becomes

$$\frac{dR}{dt} = \frac{c_{PL}(T_\infty - T_B(t))}{\mathcal{L}} \left(\frac{\alpha_L}{t}\right)^{\frac{1}{2}} \tag{7.2}$$

which is typical of the expressions for the bubble growth rate in boiling. Now consider a nucleus or bubble passing through the pump which it will do in a time of order $1/\Omega\phi$. It follows that if $\Omega\phi \gg 1/t_C$ then the bubble growth will not be inhibited by thermal

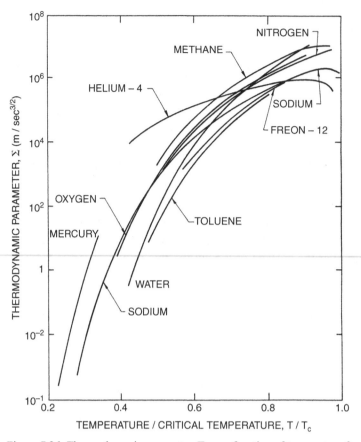

Figure 7.26. Thermodynamic parameter, Σ, as a function of temperature for various saturated fluids.

effects and explosive cavitating bubble growth will occur with the potential of causing substantial disruption of the flow and degradation of pump performance. On the other hand, if $\Omega\phi \ll 1/t_C$, most of the bubble growth will be thermally inhibited and the cavitation performance will be much improved.

To calculate t_C we need values of Σ which by its definition (equation 6.7) is a function only of the liquid temperature. Typical values of Σ for a variety of liquids are presented in figure 7.26 as a function of temperature (the ratio of temperature to critical temperature is used in order to show all the fluids on the same graph). Note that the large changes in the value of Σ are caused primarily by the change in the vapor density with temperature.

As an example, consider a cavitating flow of water in which the tension, $(p_B - p)$, is of the order of 10^4 kg/m s^2 or 0.1 bar. Then, since water at $20°C$ has a value of Σ of about 1 $m/s^{\frac{3}{2}}$, the value of t_C is of the order of 10 s. Thus, in virtually all pumps, $\Omega\phi$ will be much greater than $1/t_C$ and no thermal effect will occur. On the other hand at $100°C$, the value of Σ for water is about 10^3 $m/s^{\frac{3}{2}}$ and it follows that $t_C = 10$ μs. Thus in virtually all cases $\Omega\phi \ll 1/t_C$ and a strong thermal effect can be expected. In fact, in a given application there will exist a "critical" temperature above which one should expect a thermal effect on cavitation. For a water pump rotating at 3000 rpm this "critical" temperature is about $70°C$, a value which is consistent with the experimental measurements of pump performance.

The principal difficulty with the above approach is in finding some way to evaluate the tension, $p_B - p$, for use in equation 7.1 in order to calculate t_C. Alternatively, the experimental data could be examined for guidance in establishing a criterion based on the above model. To do so equation 7.1 is rewritten in terms of dimensionless groups as follows:

$$\Omega \phi t_C = \frac{1}{2} \left\{ \frac{p_B - p}{\frac{1}{2} \rho_L \Omega^2 R_T^2} \right\} \left\{ \frac{R_T^2 \Omega^3 \phi}{\Sigma^2} \right\} \tag{7.3}$$

where the expression in the first curly brackets on the right-hand side could be further approximated by $(-C_{pmin} - \sigma)$ where C_{pmin} is a characteristic minimum pressure coefficient. It follows that the borderline between a flow which is broken down due to cavitation in the absence of thermal effects and a flow which is not broken down due to a beneficial thermal effect occurs when the ratio of times, $\Omega \phi t_C$, takes some critical value which we will denote by β. Equation 7.3 with $\Omega \phi t_C$ set equal to β would then define a critical breakdown cavitation number, σ_x (σ_a or σ_b) as follows:

$$\sigma_x = -C_{pmin} - 2\beta \frac{\Sigma^2}{R_T^2 \Omega^3 \phi} \tag{7.4}$$

The value of σ_x in the absence of thermal effects should then be $(\sigma_x)_0 = -C_{pmin}$ and equation 7.4 can be presented in the form

$$\frac{\sigma_x}{(\sigma_x)_0} = 1 - 2\beta \frac{\Sigma^2}{R_T^2 \Omega^3 \phi (\sigma_x)_0} \tag{7.5}$$

It would then follow that the ratio of critical cavitation numbers, $\sigma_x/(\sigma_x)_0$, should be a simple function of the modified thermal effect parameter, Σ^*, defined by

$$\Sigma^* = \Sigma / \left\{ R_T^2 \Omega^3 \phi (\sigma_x)_0 \right\}^{\frac{1}{2}} \tag{7.6}$$

To test this hypothesis, data from a range of different experiments is presented in figure 7.27. It can be seen that the data correspond very roughly to some kind of common curve for all these different pumps and liquids. The solid line corresponds to equation 7.5 with an arbitrarily chosen value of $\beta = 5 \times 10^{-6}$. Consequently this attempt to model the thermal effect has succeeded to a limited degree. It should however be noted that the horizontal scatter in the data in figure 7.27 is more than a decade, though such scatter may be inevitable given the range of impeller geometries. We have also omitted one set of data, namely that of Chivers (1969), since it lies well to the left of the data included in the figure.

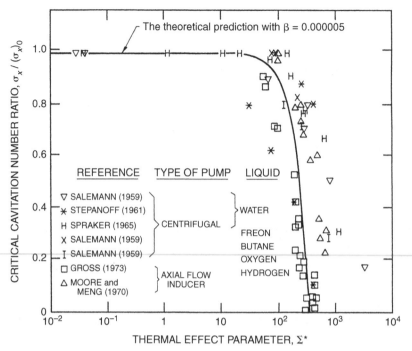

Figure 7.27. The ratio of the critical cavitation number σ_x (σ_a or σ_b) to $(\sigma_x)_0$ (the value of σ_x in the absence of any thermal effect) as a function of the thermal effect parameter, Σ^*. Data is shown for a variety of pumps and liquids.

A number of purely empirical approaches to the same problem have been suggested in the past. All these empirical methods seek to predict the change in *NPSH*, say $\Delta NPSH$, due to the thermal effect. This quantity $\Delta NPSH$ is the increment by which the cavitation performance characteristic would be shifted to the left as a result of the thermal effect. The method suggested by Stahl and Stepanoff (1956) and Stepanoff (1961, 1964) is widely used; it is based on the premise that the cavitation characteristic of a particular pump operating at a particular speed with two different liquids (or with two different temperatures in the same liquid) would be horizontally shifted by

$$\Delta NPSH = H_{T1} - H_{T2} \tag{7.7}$$

where the quantities H_{T1} and H_{T2} only depend on the thermodynamic properties of the two individual fluids considered separately. This generic property is denoted by H_T. For convenience, Stepanoff also uses the symbol B' to denote the group $\rho_L^2 c_{PL} T_\infty / \rho_V^2 \mathcal{L}^2$ which also occurs in Σ and almost all analyses of the thermal effect. Then, by examining data from a number of single-stage 3500 *rpm* pumps, Stahl and Stepanoff arrived at an empirical relation between H_T and the thermodynamic

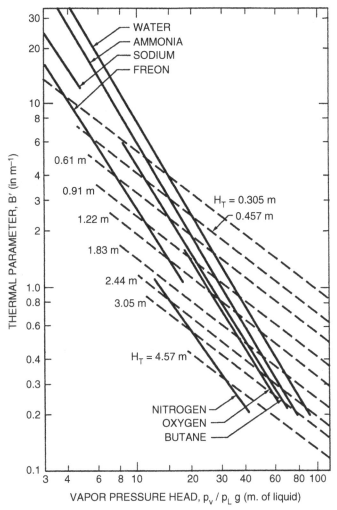

Figure 7.28. The parameter, B' (in m^{-1}), for several liquids and H_T (in m) for centrifugal pumps operating at design flow rate and at a 3% head drop (adapted from Knapp *et al.* 1970 and Stepanoff 1964).

properties of the following form:

$$H_T \text{ (in } m) = 28.9 \rho_L g / p_V (B')^{\frac{4}{3}} \tag{7.8}$$

where $p_V / g \rho_L$ is the vapor pressure head (in m) and B' is in m^{-1}. This relation is presented graphically in figure 7.28. Clearly equation 7.8 (or figure 7.28) can be used to find H_T for the desired liquid and operating temperature and for the reference liquid at the reference operating temperature. Then the cavitation performance under the desired conditions can be obtained by application of the shift given by equation 7.7 to the known cavitation performance under the reference conditions.

7.8 Free Streamline Methods

The diversity of types of cavitation in a pump and the complexity of the two-phase flow which it generates mean that reliable analytical methods for predicting the cavitating performance characteristics are virtually non-existent. However if the cavity flow can be approximated by single, fully developed or attached cavities on each blade, then this allows recourse to the methods of free streamline theory for which the reader may wish to consult the reviews by Tulin (1964) and Wu (1972) or the books by Birkhoff and Zarantonello (1957) and Brennen (1994). The analytical approaches can be subdivided into linear theories which are applicable to slender, streamlined flows (Tulin 1964) and non-linear theories which are more accurate but can be mathematically much more complex (Wu 1972). Both approaches to free streamline flows have been used in a wide range of cavity flow problems and it is necessary to restrict the present discussion to some of the solutions of relevance to attached cavitation in pumps.

It is instructive to begin by quoting some of the results obtained for single hydrofoils for which the review by Acosta (1973) provides an excellent background. In particular we will focus on the results of approximate linear theories for a partially cavitating or supercavitating flat plate hydrofoil. The partially cavitating solution (Acosta 1955) yields a lift coefficient

$$C_L = \pi\alpha\left[1 + (1-\ell)^{-\frac{1}{2}}\right] \tag{7.9}$$

where ℓ is the ratio of the cavity length to the chord of the foil and is related to the cavitation number, σ, by

$$\frac{\sigma}{2\alpha} = \frac{2 - \ell + 2(1-\ell)^{\frac{1}{2}}}{\ell^{\frac{1}{2}}(1-\ell)^{\frac{1}{2}}} \tag{7.10}$$

Thus, for a given cavity length, ℓ, and a given angle of incidence, α, the cavitation number follows from equation 7.10 and the lift coefficient from equation 7.9. Note that as $\ell \to 0$ the value of C_L tends to the theoretical value for a non-cavitating flat plate, namely $2\pi\alpha$. The corresponding solution for a supercavitating flat plate was given by Tulin (1953) in his pioneering paper on linearized cavity flows. In this case

$$C_L = \pi\alpha\ell\left[\ell^{\frac{1}{2}}(\ell-1)^{-\frac{1}{2}} - 1\right] \tag{7.11}$$

$$\alpha\left(\frac{2}{\sigma} + 1\right) = (\ell-1)^{\frac{1}{2}} \tag{7.12}$$

where now, of course, $\ell > 1$.

The lift coefficient and the cavity length from equations 7.9 to 7.12 are plotted against cavitation number in figure 7.29 for a typical angle of incidence of $\alpha = 4°$. Note that as $\sigma \to \infty$ the fully wetted lift coefficient, namely $2\pi\alpha$, is recovered from

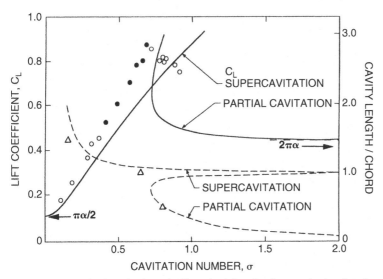

Figure 7.29. Typical results from the linearized theories for a cavitating flat plate at an angle of incidence of 4°. The lift coefficients, C_L (solid lines), and the ratios of cavity length to chord, ℓ (dashed lines), are from the supercavitation theory of Tulin (1953) and the partial cavitation theory of Acosta (1955). Also shown are the experimental results of Wade and Acosta (1966) for ℓ ($\triangle$) and for C_L (○ and ●) where the open symbols represent points of stable operation and the solid symbols denote points of unstable cavity operation.

the partial cavitation solution and that as $\sigma \to 0$ the lift coefficient tends to $\pi\alpha/2$. Notice also that both the solutions become pathological when the length of the cavity approaches the chord length ($\ell \to 1$). However, if some small portion of each curve close to $\ell = 1$ were eliminated, then the characteristic decline in the performance of the hydrofoil as the cavitation number is decreased is readily observed. It also compares well with the experimental observations as illustrated by the favorable comparison with the data of Wade and Acosta (1966) included in figure 7.29. Consequently, as the cavitation number is decreased, a single foil exhibits only a small change in the performance or lift coefficient until some critical value of σ (about 0.7 in the case of figure 7.29) is reached. Below this critical value the performance begins to "breakdown" quite rapidly. Thus, even a single foil mirrors the typical cavitation performance experienced in a pump. On a more detailed level, note that the small increase in the supercavitating lift coefficient which occurs as the cavitation number is decreased toward the critical value of σ is, in fact, observed experimentally with many single hydrofoils (for example, Wade and Acosta 1966) as well as in some pumps.

The peculiar behaviour of the analytical solutions close to the critical cavitation number is related to an instability which is observed when the cavity length is of the same order as the chord of the foil. However, we delay further discussion of this until the appropriate point in the next chapter (see section 8.10). Some additional data on the variation of the lift coefficient with angle of incidence is included in that later section.

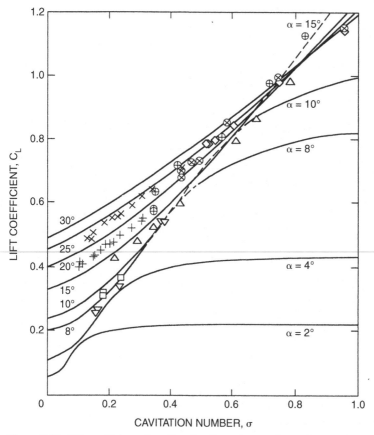

Figure 7.30. Lift coefficients for a flat plate from the non-linear theory of Wu (1962). The experimental data (Parkin 1958) is for angles of incidence as follows: 8° ($\triangledown$), 10° ($\square$), 15° ($\triangle$), 20° ($\oplus$), 25° ($\otimes$), and 30° ($\diamondsuit$). Also shown is some data of Silberman (1959) in a free jet tunnel: 20° (+) and 25° ($\times$).

Before leaving the subject of the single cavitating foil we should note that more exact, non-linear solutions for a flat plate or an arbitrarily shaped profile have been generated by Wu (1956, 1962), Mimura (1958) and others. As an example of these non-linear results, the lift and drag coefficients at various cavitation numbers and angles of incidence are presented in figures 7.30 and 7.31 where they are compared with the experimental data of Parkin (1958) and Silberman (1959). Data both for supercavitating and partially cavitating conditions are shown in these figures, the latter occurring at the higher cavitation numbers and lower incidence angles (the dashed parts of the curves represent a somewhat arbitrary smoothing through the critical region in which the cavity lengths are close to the chord length). This comparison demonstrates that the non-linear theory yields values which are in good agreement with the experimental measurements. In the case of circular-arc hydrofoils, Wu and Wang (1964) have shown similar agreement with the data of Parkin (1958) for this type of profile. For a recent treatment of supercavitating single foils the reader is referred to the work of Furuya and Acosta (1973).

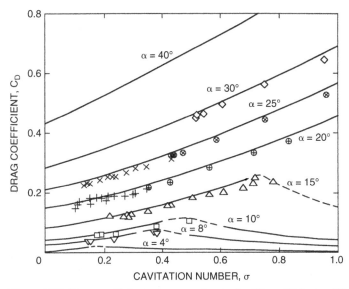

Figure 7.31. Drag coefficients corresponding to the lift coefficients of figure 7.30.

7.9 Supercavitating Cascades

We now turn to the free streamline analyses which are most pertinent to turbomachines, namely solutions and data for cavitating cascades. Both partially cavitating
and supercavitating cascades (see figure 5.10) have been analysed using free streamline methods. Clearly cavities initiated at the leading edge are more likely to extend
beyond the trailing edge when the solidity and the stagger angle are small. Such
cascade geometries are more characteristic of propellers and, therefore, the supercavitating cascade results are more often applied in that context. On the other hand, most
cavitating pumps have large solidities (> 1) and large stagger angles. Consequently,
partial cavitation is the more characteristic condition in pumps, particularly since the
pressure rise through the pump is likely to collapse the cavity before it emerges from
the blade passage. In this section we will discuss the supercavitating analyses and
data; the next section will deal with the partially cavitating results.

Free streamline methods were first applied to the problems of a cavitating cascade
by Betz and Petersohn (1931) who used a linearized method to solve the problem of
infinitely long, open cavities produced by a cascade of flat plate hydrofoils. Extensions
to this linear, supercavitating solution were generated by Sutherland and Cohen (1958)
who solved the problem of finite supercavities behind a flat plate cascade and by
Acosta (1960) who generalized this to a cascade of circular arc hydrofoils. Other early
contributions to linear cascade theory for supercavitating foils include the models of
Duller (1966) and Hsu (1972) and the inclusion of the effect of rounded leading edges
by Furuya (1974). Non-linear solutions were first obtained by Woods and Buxton
(1966) for the case of a cascade of flat plates. Later Furuya (1975) expanded this work
to include foils of arbitrary geometry.

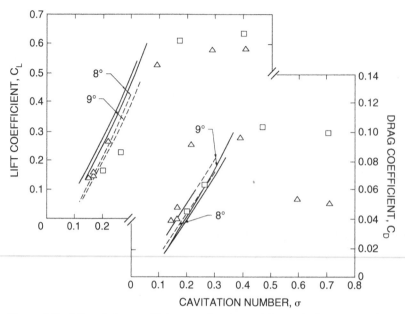

Figure 7.32. Lift and drag coefficients as functions of the cavitation number for cascades of solidity, 0.625, and blade angle, $\beta_b = 45° + \alpha$, operating at angles of incidence, α, of 8° ($\triangle$) and 9° ($\square$). The points are from the experiments of Wade and Acosta (1967) and the analytical results for a supercavitating cascade are from the linear theory of Duller (1966) (dashed lines) and the non-linear theory of Furuya (1975) (solid lines).

A substantial body of data on the performance of cavitating cascades has been accumulated through the efforts of Numachi (1961, 1964), Wade and Acosta (1967) and others. This allows comparison with the analytical models, in particular the supercavitating theories. Figure 7.32 provides such a comparison between measured lift and drag coefficients (defined as normal and parallel to the direction of the incident stream) for a particular cascade and the theoretical results from the supercavitating theories of Furuya (1975) and Duller (1966). Note that the measured lift coefficients exhibit a clear decline in cascade performance as the cavitation number is reduced and the supercavities grow. However, it is important to observe that this degradation does not occur until the cavitation is quite extensive. The cavitation inception numbers for the experiments were $\sigma_i = 2.35$ (for 8°) and $\sigma_i = 1.77$ (for 9°). However the cavitation number must be lowered to about 0.5 before the performance is adversely affected. Consequently there is a significant range of intermediate cavitation numbers within which partial cavitation is occurring and within which the performance is little changed.

For the cascades and incidence angles used in the example of figure 7.32, Furuya (1975) shows that the linear and non-linear supercavitation theories yield similar results which are close to those of the experiments. This is illustrated in figure 7.32. However, Furuya also demonstrates that there are circumstances in which the linear theories can be substantially in error and for which the non-linear results are

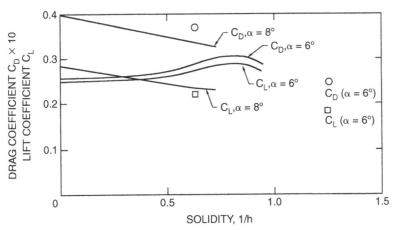

Figure 7.33. Lift and drag coefficients as functions of the solidity for cascades of blade angle, $\beta_b = 45° + \alpha$, operating at the indicated angles of incidence, α, and at a cavitation number, $\sigma = 0.18$. The points are from the experiments of Wade and Acosta (1967) and the lines are from the non-linear theory of Furuya (1975). Reproduced from Furuya (1975).

clearly needed. The effect of the solidity on the results is also important because it is a major design factor in determining the number of blades in a pump or propeller. Figure 7.33 illustrates the effect of solidity when large supercavities are present ($\sigma = 0.18$). Note that the solidity has remarkably little effect at the smaller angles of incidence.

7.10 Partially Cavitating Cascades

In the context of pumps, the solutions by Acosta and Hollander (1959) and Stripling and Acosta (1962) of partial cavitation in a semi-infinite cascade of infinitely thin blades and the solution by Wade (1967) of a finite cascade of partially cavitating foils provide a particularly valuable means of analyzing the performance of two-dimensional cascades with blade cavities. More recently the three dimensional aspects of these solutions have been explored by Furuya (1974). As a complement to purely analytical methods, more heuristic approaches are possible in which the conventional cascade analyses (see sections 3.2 and 3.5) are supplemented by lift and drag data for blades operating under cavitating conditions.

Partly for the purposes of example and partly because the results are useful, we shall recount here the results of the free-streamline solution of Brennen and Acosta (1973). This is a slightly modified version of the Acosta and Hollander solution for partial cavitation in a cascade of infinitely thin, flat blades. The modification was to add finite thickness to the blades. As we shall see, this can be important in terms of the relevance of the theory.

A sketch of the cascade geometry is shown in figure 7.34. A single parameter is introduced to the solution in order to yield finite blade thickness. This parameter implies a ratio, d, of the blade thickness far downstream to the normal spacing between

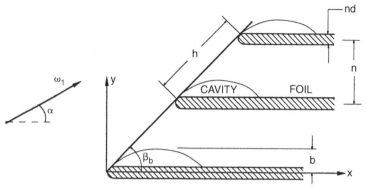

Figure 7.34. Schematic of partially cavitating cascade of flat blades of thickness nd (Brennen and Acosta 1973).

the blades. It also implies a radius of curvature of the parabolic leading edge of the blade, κ, given by

$$\frac{1}{\kappa} \approx \frac{d^2 \beta_b^3}{\pi h (1 + \sigma_c)} \tag{7.13}$$

where σ_c is the choked cavitation number (see below). Equation 7.13 and the fact that the ultimate thickness is not reached until about half a blade spacing downstream, both imply very sharp leading edges.

One of the common features of all of these free streamline solutions is that there exists a certain minimum cavitation number at which the cavity becomes infinitely long and below which there are no solutions. This minimum cavitation number is called the choked cavitation number, σ_c. Were such a flow to occur in practice, it would permit large deviation angles at discharge and a major degradation of performance. Consequently the choked cavitation number, σ_c, is often considered an approximation to the breakdown cavitation number, σ_b, for the pump flow which the cascade solution represents. The Brennen and Acosta solution yields a choked cavitation number given by

$$\sigma_c = \left[1 + 2 \sin \frac{\alpha}{2} \sec \frac{\beta_b}{2} \sin \frac{(\beta_b - \alpha)}{2} + 2d \sin^2 \frac{\beta_b}{2} \right]^2 - 1 \tag{7.14}$$

which, since the solution is only valid for small incidence angles, α, and since β_b is normally small, yields

$$\sigma_c \approx \alpha (\beta_b - \alpha) + \beta_b^2 d \tag{7.15}$$

Furthermore, at a general cavitation number, σ, the maximum thickness, b, of the cavity is given by

$$\frac{b}{n} = 2\pi \left[d - (1 + \sigma)^{\frac{1}{2}} + \sin(\beta_b - \alpha) / \sin \beta_b \right] \tag{7.16}$$

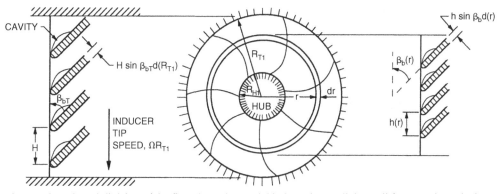

Figure 7.35. The subdivision of the flow through an axial inducer into radial annuli for cascade analysis.

or

$$\frac{b}{n} \approx 2\pi \left[d - \frac{\alpha}{\beta_b} - \frac{\sigma}{2} \right] \qquad (7.17)$$

As a rough example, consider a $10°$ helical inducer ($\beta_b = 10°$) with a fractional blade thickness of $d = 0.15$ operating at a flow coefficient, $\phi = 0.08$, so that the incidence angle, $\alpha = 4°$ (see figure 7.38). Then, according to the relation 7.15, the choked cavitation number is $\sigma_c = 0.0119$ which is close to the observed breakdown cavitation number (see figure 7.40). It is important to note the role played by the blade thickness in this typical calculation because with $d = 0$ the result is $\sigma_c = 0.0073$. Note also that with infinitely thin blades, the expression 7.15 predicts $\sigma_c = 0$ at zero incidence. Thus, the blade thickness is important in estimating the choked cavitation number in any pump.

Most pumps or inducer designs incorporate significant variations in α, β_b and d over the inlet plane and hence the above analysis has to be performed as a function of the inlet radial position as indicated in figure 7.35. Typical input data for such calculations are shown in figures 7.36, 7.38, and 7.39 for the Saturn J2 and F1 liquid oxygen turbopumps, for the $9°$ helical inducer, Impeller III, and for the SSME low pressure liquid oxygen impeller, Impeller IV. To proceed with an evaluation of the flow, the cascade at each radial annulus must then be analyzed in terms of the cavitation number, $\sigma(r)$, pertaining to that particular radius, namely

$$\sigma(r) = (p_1 - p_V)/\frac{1}{2}\rho_L \Omega^2 r^2 \qquad (7.18)$$

Specific values of this cavitation number at which choking occurs in each cascade can then be obtained from equations 7.14 or 7.15; we denote these by $\sigma_c(r)$. It follows

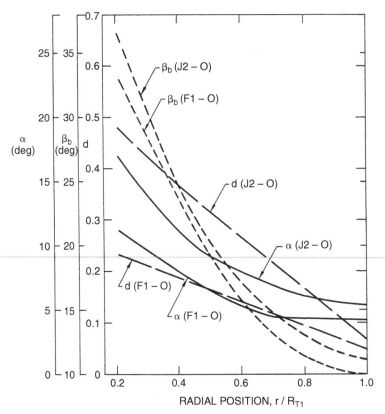

Figure 7.36. Radial variations of the blade angle, β_b, blade thickness to normal spacing ratio, d, and incidence angle α (for $\phi = 0.097$) for the oxidizer turbopumps in the Saturn J1 and F1 engines (from Brennen and Acosta 1973).

that the *overall* pump cavitation number at which the flow in each annulus will be choked is given by $\sigma_{cT}(r)$ where

$$\sigma_{cT}(r) = \sigma_c(r)r^2 / R_{T1}^2 \tag{7.19}$$

Typical data for $\sigma_{cT}(r)$ for the Saturn J2 and F1 oxidizer pumps are plotted in figure 7.37; additional examples for Impellers III and IV are shown in figure 7.40. Note that, in theory, the flow at one particular radial location will become choked before that at any other radius. The particular location will depend on the radial distributions of blade angle and blade thickness and may occur near the hub (as in the cases shown in figure 7.37) or near the tip. However, one might heuristically argue that once the flow at any radius becomes choked, the flow through the pump will reach breakdown. On this basis, the data of figure 7.37 would predict breakdown in the J2-O turbopump at $\sigma_b \approx 0.019$ and at 0.0125 for the F1-O turbopump. In table 7.2 and figure 7.37 these predictions are compared with the observed values from tests in which water is used. The agreement appears quite satisfactory. Some data obtained from tests with

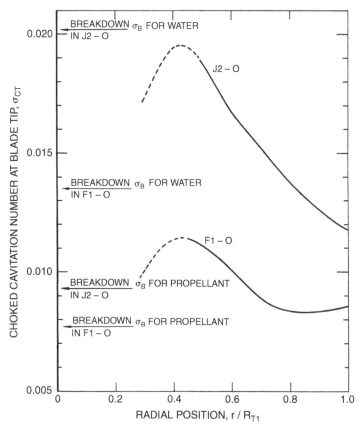

Figure 7.37. The tip cavitation numbers at which the flow at each radial location becomes choked. Data is shown for the Saturn J2 and F1 oxidizer turbopumps (see figure 7.36); experimentally observed breakdown cavitation numbers in water and propellant are also shown (from Brennen and Acosta 1973).

propellant rather than water is also shown in figure 7.37 and exhibits less satisfactory agreement; this is probably the result of thermal effects in the propellant which are not present in the water tests. Moreover, as expected, the predicted results do change with flow coefficient (since this alters the angle of incidence) as illustrated in figure 7.40.

Perhaps the most exhaustive experimental investigation of breakdown cavitation numbers for inducers is the series of experiments reported by Stripling (1962) in which inducers with blade angles at the tip, β_{bT1}, varying from 5.6° to 18°, various leading edge geometries, blade numbers of 3 and 4 and two hub-to-tip ratios were investigated. Some of Stripling's experimental data is presented in figure 7.41 where the σ_b values are plotted against the flow coefficient, ϕ_1. In his paper Stripling argues that the data correlate with the parameter $\phi_1 \sin\beta_{bT1}/(1 + \cos\beta_{bT1})$ but, in fact, the experimental data are much better correlated with ϕ_1 alone as demonstrated in figure 7.41. There is no satisfactory explanation for the fact that σ_b correlates better with ϕ_1.

Stripling correlates his data with the theoretical values of the choked cavitation number which one would obtain from the above theory in the case of infinitely thin

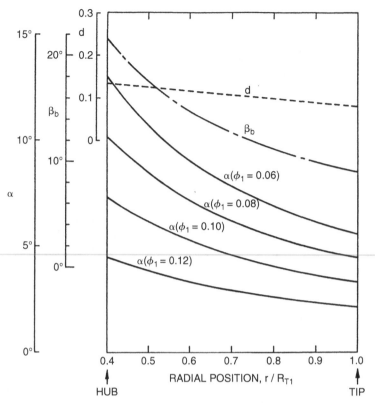

Figure 7.38. Radial variations of the blade angle, β_b, blade thickness to normal spacing ratio, d, and incidence angle, α, for the 9° helical inducer, Impeller III (from Brennen and Acosta 1976).

blades. (In this limit the expression for σ_c is more easily obtained by simultaneous solution of the Bernoulli equation and an equation for the momentum parallel with the blades as Stripling demonstrates.) More specifically, Stripling uses the blade angles, β_{b1}, and incidence angles at the rms radius, R_{RMS}, where

$$R_{RMS} = \left[\frac{1}{2} \left(R_{T1}^2 + R_{H1}^2 \right) \right]^{\frac{1}{2}} \tag{7.20}$$

His theoretical results then correspond to the dashed lines in figure 7.41. When the blade thickness term is added as in equation 7.15 the choked cavitation numbers are given by the solid lines in figure 7.41 which are considerably closer to the experimental values of σ_b than the dashed lines. The remaining discrepancy could well be due to the fact that the σ_c values are larger at some radius other than R_{RMS} and hence breakdown occurs first at that other radius.

Up to this point we have only discussed the calculation of the choked or breakdown cavitation number from the analysis of a partially cavitating cascade. There remains the issue of how to predict the degradation in the head or the cavitation head losses prior to breakdown. The problem here is that calculation of the lift from these analyses

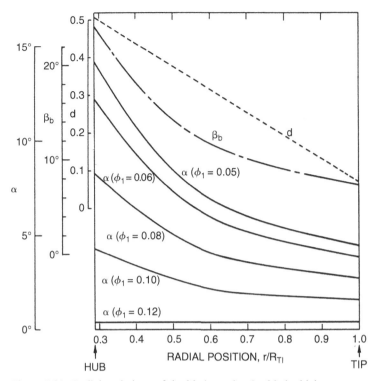

Figure 7.39. Radial variations of the blade angle, β_b, blade thickness to normal spacing ratio, d, and incidence angle, α, at inlet to the SSME low pressure liquid oxygen pump, Impeller IV (from Brennen and Acosta 1976).

Table 7.2. *Theoretical predictions of breakdown cavitation numbers compared with those observed during water tests with various inducer pumps.*

Inducer	Theory σ_c	Observed σ_b
Saturn J2 Oxidizer Inducer	0.019	0.020
Saturn F1 Oxidizer Inducer	0.012	0.013
SSME Low Pressure LOX Pump	0.011	0.012
9° Helical Impeller III	0.009	0.012

produces little for, as one could anticipate, a small partial cavity will not significantly alter the performance of a cascade of higher solidity since the discharge, with or without the cavity, is essentially constrained to follow the direction of the blades. The hydraulic losses which one seeks are additional (or possibly negative) frictional losses generated by the disruption to the flow caused by the cavitation. A number of authors, including Stripling and Acosta (1962), have employed modifications to cascade analyses in order to evaluate the loss of head, ΔH, due to cavitation. One way to view this loss is to recognize that the presence of a cavity in the blade passage causes a reduction in the cross-sectional area available to the liquid flow. When the cavity collapses this

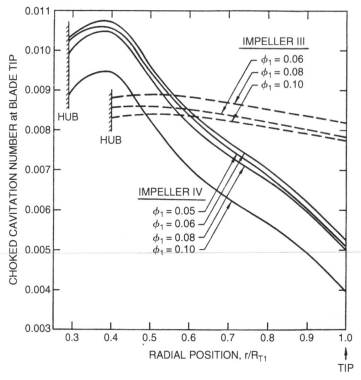

Figure 7.40. The tip cavitation numbers at which the flow at each radial location becomes choked for Impellers III and IV (see figures 7.38 and 7.39) and different flow coefficients, ϕ_1 (from Brennen and Acosta 1976).

area increases creating a "diffuser" which is not otherwise present. Hydraulic losses in this "diffuser" flow could be considered responsible for the cavitation head loss and could be derived from knowledge of the cavity blockage, b/n.

7.11 Cavitation Performance Correlations

Finally we provide brief mention of several of the purely empirical methods which are used in practice to generate estimates of the cavitation head loss in pumps. These often consist of an empirical correlation between the cavitation head loss, ΔH, the net positive suction head, $NPSH$, and the suction specific speed, S. Commonly this correlation is written as

$$\Delta H = P(S) \times NPSH \tag{7.21}$$

where the dimensionless parameter, $P(S)$, is established by prior experience. A typical function, $P(S)$, is presented in figure 7.42. Such methods can only be considered approximate; there is no fundamental reason to believe that $\Delta H / NPSH$ is a function

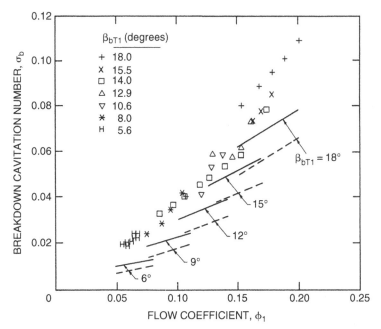

Figure 7.41. The breakdown cavitation numbers for a series of inducers by Stripling (1962) plotted against inlet flow coefficient. The inducers have inlet blade angles at the tip, β_{bT1} (in degrees), as indicated. Also shown are the results of the cascade analysis (equation 7.15) applied at the rms radius, R_{RMS}, with blade thickness (solid line) and without blade thickness (dashed line).

only of the suction specific speed, S, for all pumps though it will certainly correlate with that parameter for a given pump and a given liquid at a given Reynolds number and a given temperature. A more informed approach is to select a value of the cavitation number, σ_W, which is most fundamental to the interaction of the flow and the pump blade namely

$$\sigma_W = (p_1 - p_V)/\frac{1}{2}\rho_L w_1^2 \tag{7.22}$$

Then, using the definition of $NPSH$ (section 5.2) and the velocity triangle,

$$NPSH = \left((1+\sigma_W)v_{m1}^2 + \sigma_W\Omega^2 R_{T1}^2\right)/2g \tag{7.23}$$

It is interesting to observe that the estimate of the cavitation-free $NPSH$ for mixed flow pumps obtained empirically by Gongwer (1941) namely

$$\left(1.8v_{m1}^2 + 0.23\Omega^2 R_{T1}^2\right)/2g \tag{7.24}$$

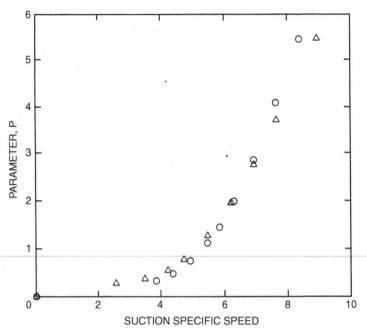

Figure 7.42. Some data on the cavitation head loss parameter, $P = \Delta H / NPSH$, for axial inducer pumps. The two symbols are for two different pumps.

and his estimate of the breakdown $NPSH$ namely

$$\left(1.49 v_{m1}^2 + 0.085 \Omega^2 R_{T1}^2\right) / 2g \tag{7.25}$$

correspond quite closely to specific values of σ_W, namely $\sigma_W \approx 0.3$ and $\sigma_W \approx 0.1$ respectively.

8

Pump Vibration

8.1 Introduction

The trend toward higher speed, high power density liquid turbomachinery has inevitably increased the potential for fluid/structure interaction problems, and the severity of those problems. Even in the absence of cavitation and its complications, these fluid structure interaction phenomena can lead to increased wear and, under the worst conditions, to structural failure. Exemplifying this trend, the Electrical Power Research Institute (Makay and Szamody 1978) has recognized that the occurrence of these problems in boiler feed pumps has contributed significantly to downtime in conventional power plants.

Unlike the cavitation issues, unsteady flow problems in liquid turbomachines do not have a long history of research. In some ways this is ironic since, as pointed out by Ek (1957) and Dean (1959), the flow within a turbomachine must necessarily be unsteady if work is to be done on or by the fluid. Yet many of the classical texts on pumps or turbines barely make mention of unsteady flow phenomena or of design considerations that might avoid such problems. In contrast to liquid turbomachinery, the literature on unsteady flow problems in gas turbomachinery is considerably more extensive, and there are a number of review papers that provide a good survey of the subject (for example, McCroskey 1977, Cumpsty 1977, Mikolajczak *et al.* 1975, Platzer 1978, Greitzer 1981). We will not attempt a review of this literature but we will try, where appropriate, to indicate areas of useful cross-reference. It is also clear that this subject incorporates a variety of problems ranging, for example, from blade flutter to fluid-induced rotordynamic instability. Because of this variety and the recent vintage of the fundamental research, no clear classification system for these problems has yet evolved and there may indeed be some phenomena that have yet to be properly identified. It follows that the classification system that we will attempt here will be tentative, and not necessarily comprehensive. Nevertheless, it seems that three different categories of flow oscillation can occur, and that there are a number phenomena within each of the three categories. We briefly list them here and return to some in the sections that follow.

[A] **Global Flow Oscillations.** A number of the identified vibration problems involve large scale oscillations of the flow. Specific examples are:

[A1] Rotating stall or rotating cavitation occurs when a turbomachine is required to operate at a high incidence angle close to the value at which the blades may stall. It is often the case that stall will first be manifest on a small number of adjacent blades and that this "stall cell" will propagate circumferentially at some fraction of the main impeller rotation speed. This phenomenon is called rotating stall and is usually associated with turbomachines having a substantial number of blades (such as compressors). It has, however, also been reported in centrifugal pumps. When the turbomachine cavitates the same phenomenon may still occur, perhaps in some slightly altered form. Such circumstances will be referred to as "rotating stall with cavitation." But there is also a different phenomenon which can occur in which one or two blades manifest a greater degree of cavitation and this "cell" propagates around the rotor in a manner superficially similar to the propagation of rotating stall. This phenomenon is known as "rotating cavitation."

[A2] Surge is manifest in a turbomachine that is required to operate under highly loaded circumstances where the slope of the head rise/flow rate curve is positive. It is a system instability to which the dynamics of all the components of the system (reservoirs, valves, inlet and suction lines and turbomachine) contribute. It results in pressure and flow rate oscillations throughout the system. When cavitation is present the phenomenon is termed "auto-oscillation" and can occur even when the slope of the head rise/flow rate curve is negative.

[A3] Partial cavitation or supercavitation can become unstable when the length of the cavity approaches the length of the blade so that the cavity collapses in the region of the trailing edge. Such a circumstance can lead to violent oscillations in which the cavity length oscillates dramatically.

[A4] Line resonance occurs when one of the blade passing frequencies in a turbomachine happens to coincide with one of the acoustic modes of the inlet or discharge line. The pressure oscillation magnitudes associated with these resonances can often cause substantial damage.

[A5] It has been speculated that an axial balance resonance could occur if the turbomachine is fitted with a balance piston (designed to equalize the axial forces acting on the impeller) and if the resonant frequency of the balance piston system corresponds with the rotating speed or some blade passing frequency. Though there exist several apocryphal accounts of such resonances, the phenomenon has yet to be documented experimentally.

[A6] Cavitation noise can sometimes reach a sufficient amplitude to cause resonance with structural frequencies of vibration.

[A7] The above items all assume that the turbomachine is fixed in a non-accelerating reference frame. When this is not the case the dynamics of the turbomachine may play a crucial role in generating an instability that involves the vibration of that machine as a whole. Such phenomena, of which the Pogo instabilities are, perhaps, the best documented examples, are described further in section 8.13.

[B] **Local Flow Oscillations.** Several other vibration problems involve more localized flow oscillations and vibration of the blades:

[B1] Blade flutter. As in the case of airfoils, there are circumstances under which an individual blade may begin to flutter (or diverge) as a consequence of the particular flow condition (incidence angle, velocity), the stiffness of the blade, and its method of support.

[B2] Blade excitation due to rotor-stator interaction. While [B1] would occur in the absence of excitation it is also true that there are a number of possible mechanisms of excitation in a turbomachine that can cause significant blade vibration. This is particularly true for a row of stator blades operating just downstream of a row of impeller blades or vice versa. The wakes from the upstream blades can cause a serious vibration problem for the downstream blades at blade passing frequency or some multiple thereof. Non-axisymmetry in the inlet, the volute, or housing can also cause excitation of impeller blades at the impeller rotation frequency.

[B3] Blade excitation due to vortex shedding or cavitation oscillations. In addition to the excitation of [B2], it is also possible that vortex shedding or the oscillations of cavitation could provide the excitation for blade vibrations.

[C] **Radial and Rotordynamic Forces.** Global forces perpendicular to the axis of rotation can generate several types of problem:

[C1] Radial forces are forces perpendicular to the axis of rotation caused by circumferential nonuniformities in the inlet flow, casing, or volute. While these may be stationary in the frame of the pump housing, the loads that act on the impeller and, therefore, the bearings can be sufficient to create wear, vibration, and even failure of the bearings.

[C2] Fluid-induced rotordynamic forces occur as the result of movement of the axis of rotation of the impeller-shaft system of the turbomachine. Contributions to these rotordynamic forces can arise from the seals, the flow through the impeller, leakage flows, or the flows in the bearings themselves. Sometimes these forces can cause a reduction in the critical speeds of the shaft system, and therefore an unforeseen limitation to its operating range. One of the common characteristics of a fluid-induced rotordynamic problem is that it often occurs at subsynchronous frequency.

Two of the subjects included in this list have a sufficiently voluminous literature to merit separate chapters. Consequently, chapter 10 is devoted to radial and rotordynamic forces, and chapter 9 to the subject of system dynamic analysis and instabilities. The remainder of this chapter will briefly describe some of the other unsteady problems encountered in liquid turbomachines.

Before leaving the issue of classification, it is important to emphasize that many of the phenomena that cause serious vibration problems in turbomachines involve an interaction between two or more of the above mentioned items. Perhaps the most widely recognized of these resonance problems is that involving an interaction between blade passage excitation frequencies and acoustic modes of the suction or discharge lines. But the literature contains other examples. For instance, Dussourd (1968) describes flow oscillations which involve the interaction of rotating stall and acoustic line frequencies. Another example is given by Marscher (1988) who investigated a resonance between the rotordynamic motions of the shaft and the subsynchronous unsteady flows associated with flow recirculation at the inlet to a centrifugal impeller.

8.2 Frequencies of Oscillation

One of the diagnostics which is often, but not always, useful in addressing a turbomachine vibration problem is to examine the dominant frequencies and to investigate how they change with rotating speed. Table 8.1 is intended as a rough guide to the kinds of frequencies at which the above problems occur. We have attempted to place the phenomena in rough order of increasing frequency partly in order to illustrate the fact that the frequencies can range all the way from a few Hz up to tens of kHz. Some of the phenomena scale with the impeller rotating speed, Ω. Others, such as surge, may vary somewhat with Ω but not linearly; still others, like cavitation noise, will be largely independent of Ω.

Of the frequencies listed in table 8.1, the blade passing frequencies need some further clarification. We will denote the numbers of blades on an adjacent rotor and stator by Z_R and Z_S, respectively. Then the fundamental blade passage frequency in so far as a single stator blade is concerned is $Z_R \Omega$ since that stator blade will experience the passage of Z_R rotor blades each revolution of the rotor. Consequently, this will represent the fundamental frequency of blade passage excitation insofar as the inlet or discharge lines or the static structure is concerned. Correspondingly, $Z_S \Omega$ is the fundamental frequency of blade passage excitation insofar as the rotor blades (or the impeller structure) are concerned. However, the excitation is not quite as simple as this for both harmonics and subharmonics of these fundamental frequencies can often be important. Note first that, while the phenomenon is periodic, it is not neccessarily sinusoidal, and therefore the excitation will contain higher harmonics, $m Z_R \Omega$ and $m Z_S \Omega$ where m is an integer. But more importantly, when the integers Z_R and Z_S have a common factor, say Z_{CF}, then, in the framework of the stator, a particular pattern of excitation is repeated at the subharmonic, $Z_R \Omega / Z_{CF}$, of the fundamental, $Z_R \Omega$. Correspondingly, in the framework of the rotor, the structure experiences subharmonic excitation at $Z_S \Omega / Z_{CF}$. These subharmonic frequencies

Table 8.1. *Typical frequency ranges of pump vibration problems*

Vibration Category	Frequency Range
A2 Surge	System dependent, 3–10 Hz in compressors
A2 Auto-oscillation	System dependent, $0.1\Omega - 0.4\Omega$
A1 Rotor rotating stall	$0.5\Omega - 0.7\Omega$
A1 Vaneless diffuser stall	$0.05\Omega - 0.25\Omega$
A1 Rotating cavitation	$1.1\Omega - 1.2\Omega$
A3 Partial cavitation oscillation	$< \Omega$
C1 Excessive radial force	Some fraction of Ω
C2 Rotordynamic vibration	Fraction of Ω when critical speed is approached.
A4 Blade passing excitation (or B2)	$Z_R\Omega/Z_{CF}, Z_R\Omega, mZ_R\Omega$ (in stator frame) $Z_S\Omega/Z_{CF}, Z_S\Omega, mZ_S\Omega$ (in rotor frame)
B1 Blade flutter	Natural frequencies of blade in liquid
B3 Vortex shedding	Frequency of vortex shedding
A6 Cavitation noise	$1 - 20\, kHz$

can be more of a problem than the fundamental blade passage frequencies because the fluid and structural damping is smaller for these lower frequencies. Consequently, turbomachines are frequently designed with values of Z_R and Z_S which have no common factors, in order to eliminate subharmonic excitation. Further discussion of blade passage excitation frequencies is included in section 8.8.

Before proceeding to a discussion of the specific vibrational problems outlined above, it may be valuable to illustrate the spectral content of the shaft vibration of a typical centrifugal pump in normal, nominally steady operation. Figure 8.1 presents examples of the spectra (for two frequency ranges) taken from the shaft of the five-bladed centrifugal Impeller X operating in the vaneless Volute A (no stator blades) at 300 rpm (5 Hz). Clearly the synchronous vibration at the shaft fundamental of 5 Hz dominates the low frequencies; this excitation may be caused by mechanical imperfections in the shaft such as an imbalance or by circumferential nonuniformities in the flow such as might be generated by the volute. It is also clear that the most dominant harmonic of shaft frequency occurs at 5Ω because there are 5 impeller blades. Note, however, that there are noticeable peaks at 2Ω and 3Ω arising from significantly nonsinusoidal excitation at the shaft frequency, Ω. The other dominant peaks labeled $1 \rightarrow 4$ represent structural resonant frequencies unaffected by shaft rotational speed.

At higher rotational speeds, more coincidence with structural frequencies occurs and the spectra contain more noise. However, interesting features can still be discerned. Figure 8.2 presents examples, taken from Miskovish and Brennen (1992), of the

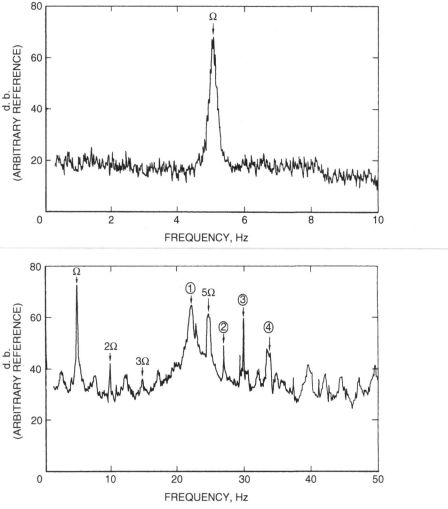

Figure 8.1. Typical spectra of vibration for a centrifugal pump (Impeller X/Volute A) operating at 300 rpm (Chamieh et al. 1985).

spectra for all six shaft forces and moments as measured in the rotating frame of Impeller X by the balance onto which that impeller was mounted. F_1, F_2 are the two rotating radial forces, M_1, M_2 are the corresponding bending moments, F_3 is the thrust, and M_3 is the torque. In this example, the shaft speed is $3000\,rpm$ ($\Omega = 100\pi\,rad/sec$) and the impeller is also being whirled at a frequency, $\omega = I\Omega/J$, where $I/J = 3/10$. Note that there is a strong peak in all the forces and moments at the shaft frequency, Ω, because of the steady radial forces caused by volute asymmetry. Rotordynamic forces would be manifest in this rotating frame at the beat frequencies $(J \pm I)\Omega/J$; note that the predominant rotordynamic effect occurs at the lower of these beat frequencies, $(J - I)\Omega/J$. The moments M_1 and M_2 are noisy because the line of action of the forces F_1 and F_2 is close to the chosen axial location of the origin of the coordinate

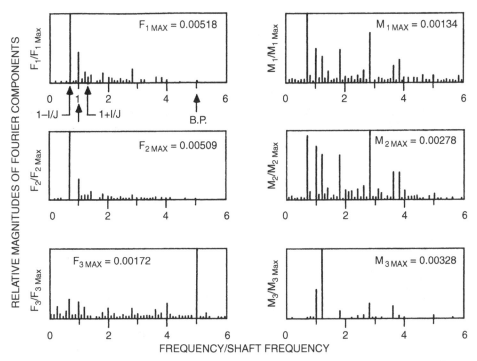

Figure 8.2. Typical frequency content of $F_1, F_2, F_3, M_1, M_2, M_3$ for Impeller X/Volute A for tests at 3000 rpm, $\phi = 0.092$, and $I/J = 3/10$. Note the harmonics $\Omega, (J \pm I)\Omega/J$ and the blade passing frequency, 5Ω.

system, the mid-point of the impeller discharge. Consequently, the magnitudes of the moments are small. One of the more surprising features in this data is the fact that the unsteady thrust contains a significant component at the blade passing frequency, 5Ω. Miskovish and Brennen (1992) indicate that the magnitude of this unsteady thrust is about $0.2 \rightarrow 0.5\%$ of the steady thrust and that the peaks occur close to the times when blades pass the volute cutwater. While this magnitude may not seem large, it could give rise to significant axial vibration at the blade passing frequency in some applications.

8.3 Unsteady Flows

Many of the phenomena listed in section 8.1 require some knowledge of the unsteady flows corresponding to the steady cascade flows discussed in sections 3.2 and 3.5. In the case of non-cavitating axial cascades, a large volume of literature has been generated in the context of gas turbine engines, and there exist a number of extensive reviews including those by Woods (1961), McCroskey (1977), Mikolajczak *et al.* (1975) and Platzer (1978). Much of the analytical work utilizes linear cascade theory, for example, Kemp and Sears (1955), Woods (1955), Schorr and Reddy (1971), and Kemp and Ohashi (1975). Some of this has been applied to the analysis of unsteady flows in pumps and extended to cover the case of radial or mixed flow machines. For example, Tsukamoto and Ohashi (1982) utilized these methods to model the start-up transients in centifugal pumps and Tsujimoto *et al.* (1986) extended the analysis to evaluate the unsteady torque in mixed flow machines.

However, most of the available methods are restricted to lightly loaded cascades and impellers at low angles of incidence. Other, more complex, theories (for example, Adamczyk 1975) are needed at larger angles of incidence and for highly cambered cascades when there is a strong coupling between the steady and unsteady flow (Platzer 1978). Moreover, most of the early theories were only applicable to globally uniform unsteady flows in which the blades all move in unison. Samoylovich (1962) appears to have been the first to consider oscillations with arbitrary interblade phase differences, the kind of analysis needed for flutter investigations (see below).

When the incidence angles are large so that the blades stall, one must resort to unsteady free streamline methods in order to model the flows (Woods 1961). Apart from the work of Sisto (1967), very little analytical work has been done on this problem which is of considerable importance in the context of turbomachinery. One of the fluid mechanical complexities is the unsteady or dynamic response of a separated flow that may lead to significant departures from the sucession of events one might construct based on a quasistatic approach. Some progress has been made in understanding the "dynamic stall" for a single foil (see, for example, Ham 1968). However, it would appear that more work is needed to understand the complex dynamic stall phenomena in turbomachines.

Unsteady free streamline analyses can be more confidentally applied to the analysis of cavitating cascades because the cavity or free streamline pressure is usually known and constant whereas the corresponding pressure for the wake flows may be varying with time in a way that is difficult to predict. Thus, for example, the unsteady response for a single supercavitating foil (Woods 1957, Martin 1962, Parkin 1962) has been compared with experimental measurements by Acosta and DeLong (1971). As an example, we present (figure 8.3) some data from Acosta and DeLong on the unsteady forces on a single foil undergoing heave oscillations at various reduced frequencies, $\omega^* = \omega c/2U$. The oscillating heave motion, d, is represented by

$$d = Re\left\{\tilde{d}e^{j\omega t}\right\} \qquad (8.1)$$

where the complex quantity, $\tilde{d}$, contains the amplitude and phase of the displacement. The resulting lift coefficient, C_L, is decomposed (using the notation of the next chapter) into

$$C_L = \bar{C}_L + Re\left\{\tilde{C}_{Lh}e^{j\omega t}\right\} \qquad (8.2)$$

and the real and imaginary parts of $\tilde{C}_{Lh}/\omega*$ which are plotted in figure 8.3 represent the unsteady lift characteristics of the foil. It is particularly important to note that

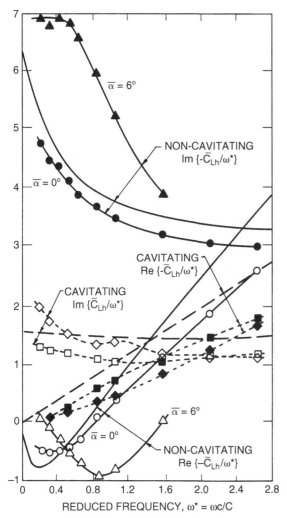

Figure 8.3. Fluctuating lift coefficient, $\tilde{C}_{Lh}$, for foils undergoing heave oscillations at a reduced frequency, $\omega^* = \omega c / U$. Real and imaginary parts of $\tilde{C}_{Lh}/\omega^*$ are presented for (a) non-cavitating flow at mean incidence angles of 0° and 6° (b) cavitating data for a mean incidence of 8°, for very long choked cavities (□) and for cavities 3 chords in length (◇). Adapted from Acosta and DeLong (1971).

substantial departures from quasistatic behavior occur for reduced frequencies as low as 0.2, though these departures are more significant in the noncavitating flow than in the cavitating flow. The lines without points in figure 8.3 present results for the corresponding linear theories and we observe that the agreement between the theory and the experiments is fairly good. Notice also that the $Re\{-\tilde{C}_{Lh}\}$ for noncavitating foils is negative at low frequencies but becomes positive at larger ω whereas the values in the cavitating case are all positive. Similar data for cavitating cascades would be necessary in order to analyse the potential for instability in cavitating, axial flow pumps. The author is not aware of any such data or analysis.

The information is similarly meagre for all of the corresponding dynamic characteristics of radial rather than axial cascades and, consequently, our ability to model dynamic instabilities in centrifugal pumps is very limited indeed.

8.4 Rotating Stall

Rotating stall is a phenomenon which may occur in a cascade of blades when these are required to operate at a high angle of incidence close to that at which the blades will stall. In a pump this usually implies that the flow rate has been reduced to a point close to or below the maximum in the head characteristic (see, for example, figures 7.5 and 7.6). Emmons *et al.* (1955) first provided a coherent explanation of propagating stall. The cascade in figure 8.4 will represent a set of blades (a rotor or a stator) operating at a high angle of incidence. Then, if blade B were stalled, this generates a separated wake and therefore increased blockage to the flow in the passage between blades B and A. This, in turn, would tend to deflect the flow away from this blockage in the manner indicated in the figure. The result would be an increase in the angle of incidence on blade A and a decrease in the angle of incidence on blade C. Thus, blade A would tend to stall while any stall on blade C would tend to diminish. Consequently, the stall "cell" would tend to move upwards in the figure or in a direction away from the oncoming flow. Of course, the stall cell could consist of a larger number of blades with more than one exhibiting increased separation or stall. The stall cell will rotate around the axis and hence the name "rotating stall." Moreover, the speed of propagation will clearly be some fraction of the circumferential component of the relative velocity, either $v_{\theta 1}$ in the case of a stator or $w_{\theta 1}$ in the case of a rotor. Consequently, in the case of a rotor, the stall rotates in the same direction as the rotor but with 50-70% of the rotor angular velocity.

In distinguishing between rotating stall and surge, it is important to note that the disturbance depicted in figure 8.4 does not necessarily imply any oscillation in the total mass flow rate through the turbomachine. Rather it implies a redistribution of that flow. On the other hand, it is always possible that the perturbation caused by rotating stall could resonate with, say, one of the acoustic modes in the inlet or discharge lines, in which case significant oscillation of the mass flow rate could occur.

While rotating stall can occur in any turbomachine, it is most frequently observed and most widely studied in compressors with large numbers of blades. Excellent reviews of this literature have been published by Emmons *et al.* (1959) and more recently by Greitzer (1981). Both point to a body of work designed to predict both the onset and consequences of rotating stall. A useful approximate criterion is that rotating stall in the rotor occurs when one approaches a maximum in the total head rise as the flow coefficient decreases. This is, however, no more than a crude approximation and Greitzer (1981) quotes a number of cases in which rotating stall occurs while the slope of the performance curve is still negative. A more sophisticated criterion that is widely used is due to Leiblein (1965), and involves the diffusion factor, Df, defined

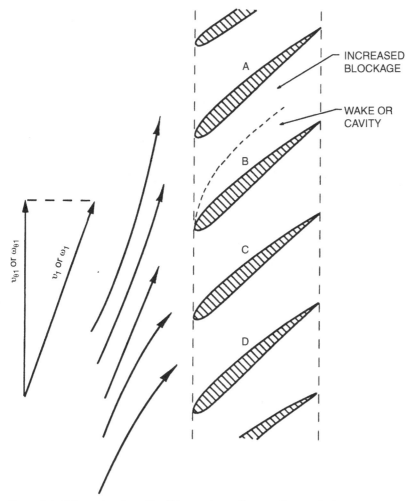

Figure 8.4. Schematic of a stall cell in rotating stall or rotating cavitation.

previously in equation 3.20. Experience indicates that rotating stall may begin when Df is increased to a value of about 0.6.

Though most of the observations of rotating stall have been made for axial compressors, Murai (1968) observed and investigated the phenomenon in a typical axial flow pump with 18 blades, a hub/tip radius ratio of 0.7, a tip solidity of 1.15, and a tip blade angle of 20°. His data on the rotating speed of the stall cell are reproduced in figure 8.5. Note that the onset of the rotating stall phenomenon occurs when the flow rate is reduced to a point below the maximum in the head characteristic. Notice also that the stall cell propagation velocities have typical values between 0.45 and 0.6 of the rotating speed. Rotating stall has not, however, been reported in pumps with a small number of blades perhaps because Df will not approach 0.6 for typical axial pumps or inducers with a small number of blades. Most of the stability theories (for example, Emmons *et al.* 1959) are based on actuator disc models of the rotor in which

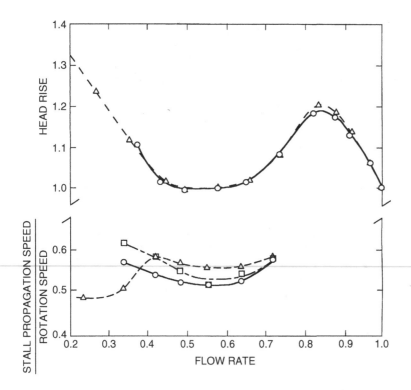

Figure 8.5. The head characteristic for an 18-bladed axial flow pump along with the measurements of the propagation velocity of the rotating stall cell relative to the shaft speed. Adapted from Murai (1968). Data is shown for three different inlet pressures. Flow and head scales are dimensionless.

it is assumed that the stall cell is much longer than the distance between the blades. Such an assumption would not be appropriate in an axial flow pump with three or four blades.

Murai (1968) also examined the effect of limited cavitation on the rotating stall phenomenon and observed that the cavitation did cause some alteration in the propagation speed as illustrated by the changes with inlet pressure seen in figure 8.5. It is, however, important to emphasize the difference between the phenomenon observed by Murai in which cavitation is secondary to the rotating stall and the phenomenon to be discussed below, namely rotating cavitation, which occurs at a point on the head-flow characteristic at which the slope is negative and stable, and at which rotating stall would not occur.

Turning now to centrifugal pumps, there have been a number of studies in which rotating stall has been observed either in the impeller or in the diffuser/volute. Hergt and Benner (1968) observed rotating stall in a vaned diffuser and conclude that it only occurs with some particular diffuser geometries. Lenneman and Howard (1970) examined the blade passage flow patterns associated with rotating stall and present data on the ratio, Ω_{RS}/Ω, of the propagation velocity of the stall cell to the impeller speed, Ω. They observed ratios ranging from 0.54 to 0.68 with, typically, lower values of the ratio at lower impeller speeds and at higher flow coefficients.

Perhaps the most detailed study is the recent research of Yoshida *et al.* (1991) who made the following observations on a 7-bladed centrifugal impeller operating with a variety of diffusers, with and without vanes. Rotating stall with a single cell was observed to occur in the impeller below a certain critical flow coefficient which depended on the diffuser geometry. In the absence of a diffuser, the cell speed was about 80–90% of the impeller rotating speed; with diffuser vanes, this cell speed was reduced to the range 50–80%. When impeller rotating stall was present, they also detected the presence of some propagating disturbances with 2, 3, and 4 cells rather than one. These are probably due to nonlinearities and an interaction with blade passage excitation. Rotating stall was also observed to occur in the vaned diffuser with a speed less than 10% of the impeller speed. It was most evident when the clearance between the impeller and diffuser vanes was large. As this clearance was decreased, the diffuser rotating stall tended to disappear.

Even in the absence of blades, it is possible for a diffuser or volute to exhibit a propagating rotating "stall." Jansen (1964) and van der Braembussche (1982) first described this flow instability and indicate that the flow pattern propagates with a speed in the range of 5–25% of the impeller speed. Yoshida *et al.* (1991) observed a four-cell rotating stall in their vaneless diffusers over a large range of flow coefficients and measured its velocity as about 20% of the impeller speed.

Finally, we note that rotating stall may resonate with an acoustic mode of the inlet or discharging piping to produce a serious pulsation problem. Dussourd (1968) identified such a problem in a boiler feed system in which the rotating stall frequency was in the range $0.15\Omega \rightarrow 0.25\Omega$, much lower than usual. He also made use of the frequency domain methods of chapter 9 in modeling the dynamics of this multistage centrifugal pump system. This represents a good example of one of the many hybrid problems that can arise in systems with turbomachines.

8.5 Rotating Cavitation

Inducers or impellers in pumps that do not show any sign of rotating stall while operating under noncavitating conditions may exhibit a superficially similar phenemenon known as "rotating cavitation" when they are required to operate at low cavitation numbers. However, it is important to emphasize the fundamental difference in the two phenomena. Rotating stall occurs at locations along the head-flow characteristic at which the blades may stall, usually at flow rates for which the slope of the head/flow characteristic is positive and therefore unstable in the sense discussed in the next section. On the other hand, rotating cavitation is observed to occur at locations where the slope is negative. These would normally be considered stable operating points in the absence of cavitation. Consequently, the dynamics of the cavitation are essential

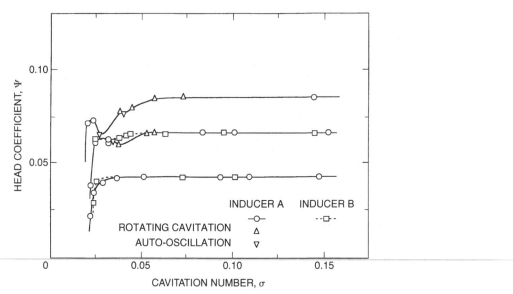

Figure 8.6. Occurrence of rotating cavitation and auto-oscillation in the performance of the cavitating inducer tested by Kamijo *et al.* (1977).

to rotating cavitation. Another difference between the phenomena is the difference in the propagating speeds.

Rotating cavitation was first explicitly identified by Kamijo, Shimura and Watanabe (1977) (see also 1980), though some evidence of it can be seen in the shaft vibration measurements of Rosemann (1965). When it has been observed, rotating cavitation generally occurs when the cavitation number, σ, is reduced to a value at which the head is beginning to be affected by the cavitation as seen in figure 8.6 taken from Kamijo *et al.* (1977). Rosenmann (1965) reported that the vibrations (that we now recognize as rotating cavitation) occurred for cavitation numbers between two and three times the breakdown value and were particularly evident at the lower flow coefficients at which the inducer was more heavily loaded.

Usually, further reduction of σ below the value at which rotating cavitation occurs will lead to auto-oscillation or surge (see below and figure 8.6). It is not at all clear why some inducers and impellers do not exhibit rotating cavitation at all but proceed directly to auto-oscillation if that instability is going to occur.

Unlike rotating stall whose rotational velocities are less than that of the rotor, rotating cavitation is characterized by a propagating velocity that is slightly larger than the impeller speed. Kamijo *et al.* (1977) (see also Kamijo *et al.* 1992) observed propagating velocities $\Omega_{RC}/\Omega \approx 1.15$, and this is very similar to one of the somewhat ambigous propagating disturbance velocities of 1.1Ω reported by Rosemann (1965).

Recently, Tsujimoto *et al.* (1992) have utilized the methods of chapter 9 to model the dynamics of rotating cavitation. They have shown that the cavitation compliance and mass flow gain factor (see section 9.14) play a crucial role in determining the instability of rotating cavitation in much the same way as these parameters influence the stability of an entire system which includes a cavitating pump (see section 8.7). Also note that the analysis of Tsujimoto *et al.* (1992) predicts supersynchronous propagating speeds in the range $\Omega_{RC}/\Omega = 1.1$ to 1.4, consistent with the experimental observations.

8.6 Surge

Surge and auto-oscillation (see next section) are system instabilities that involve not just the characteristics of the pump but those of the rest of the pumping system. They result in pressure and flow rate oscillations that can not only generate excessive vibration and reduce performance but also threaten the structural integrity of the turbomachine or other components of the system. In chapter 9 we provide more detail on general analytical approaches to this class of system instabilities. But for present purposes, it is useful to provide a brief outline of some of the characteristics of these system instabilities. To do so, consider first figure 8.7(a) in which the steady-state characteristic of the pump (head *rise* versus mass flow rate) is plotted together with the steady-state characteristic of the rest of the system to which the pump is connected (head *drop* versus mass flow rate). In steady-state operation the head rise across the pump must equal the head drop for the rest of the system, and the flow rates must be the same so that the combination will operate at the intersection, O. Consider, now, the response to a small decrease in the flow to a value just below this equilibrium point, O. Pump A will then produce more head than the head drop in the rest of the system, and this discrepancy will cause the flow rate to increase, causing a return to the equilibrium point. Therefore, because the slope of the characteristic of Pump A is less than the slope of the characteristic of the rest of the system, the point O represents a quasistatically stable operating point. On the other hand, the system with Pump B is quasistatically unstable. Perhaps the best known example of this kind of instability occurs in multistage compressors in which the characteristics generally take the shape

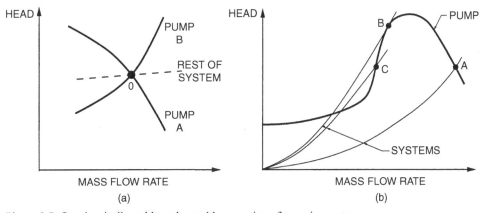

Figure 8.7. Quasistatically stable and unstable operation of pumping systems.

shown in figure 8.7(b). It follows that the operating point A is stable, point B is neutrally stable, and point C is unstable. The result of the instability at points such as C is the oscillation in the pressure and flow rate known as "compressor surge."

While the above description of quasistatic stability may help in visualizing the phenomenon, it constitutes a rather artificial separation of the total system into a "pump" and "the rest of the system." A more general analytical perspective is obtained by defining a resistance, R_i^*, for each of the series components of the system (one of which would be the pump) distinguished by the subscript, i:

$$R_i^* = \frac{d(\Delta H)}{dm} \tag{8.3}$$

where ΔH is the quasistatic head drop across that component (inlet head minus discharge head) and is a function of the mass flow rate, m. By this definition, the slope of the pump characteristic in figure 8.7(a) is $-R_{PUMP}^*$, and the slope of the characteristic of the rest of the system is R_{SYSTEM}^*. It follows that the earlier established criterion for stability is equivalent to

$$\sum_i R_i^* > 0 \tag{8.4}$$

In other words, the system is quasistatically stable if the total system resistance is positive.

Perhaps the most satisfactory interpretation of the above formulation is in terms of the energy balance of the total system. The net flux of energy out of each of the elements of the system is $m(\Delta H)_i$. Consequently, the net energy flux out of the system is

$$m \sum_i (\Delta H)_i = 0 \tag{8.5}$$

which is zero at a steady state operating point.

Suppose the stability of the system is now explored by inserting somewhere in the system a hypothetical perturbing device which causes an increase in the flow rate by dm. Then the new net energy flux out of the system, E^*, is given by

$$E^* = dm \left[\sum_i (\Delta H_i) + m \frac{d \sum_i (\Delta H_i)}{dm} \right] \tag{8.6}$$

$$= m \, dm \sum_i R_i^* \tag{8.7}$$

where the relations 8.3 and 8.5 have been used. The quantity E^* could be interpreted as the energy flux that would have to be supplied to the system through the hypothetical device in order to reestablish equilibrium. Clearly, then, if the required energy flux, E^*, is positive, the original system is stable. Therefore the criterion 8.4 is the correct condition for stability.

All of the above is predicated on the changes to the system being sufficiently slow for the pump and the system to follow the steady state operating curves. Thus the analysis is only applicable to those instabilities whose frequencies are low enough to lie within some quasistatic range. At higher frequency, it is necessary to include the inertia and compressibility of the various components of the flow. Greitzer (1976) (see also 1981) has developed such models for the prediction of both surge and rotating stall in axial flow compressors.

It is important to observe that, while quasisteady instabilities will certainly occur when $\sum_i R_i^* < 0$, there may be other dynamic instabilities that occur even when the system is quasistatically stable. One way to view this possibility is to recognize that the resistance of any flow is frequently a complex function of frequency once a certain quasisteady frequency has been exceeded. Consequently, the resistances, R_i^*, may be different at frequencies above the quasistatic limit. It follows that there may be operating points at which the total dynamic resistance over some range of frequencies is negative. Then the system would be dynamically unstable even though it may still be quasistatically stable. Such a description of dynamic instability is instructive but overly simplistic and a more systematic approach to this issue must await the methodologies of chapter 9.

8.7 Auto-Oscillation

In many installations involving a pump that cavitates, violent oscillations in the pressure and flow rate in the entire system occur when the cavitation number is decreased to values at which the head rise across the pump begins to be affected (Braisted and Brennen 1980, Kamijo *et al.* 1977, Sack and Nottage 1965, Natanzon *et al.* 1974, Miller and Gross 1967, Hobson and Marshall 1979). These oscillations can also cause substantial radial forces on the shaft of the order of 20% of the axial thrust (Rosenmann 1965). This surge phenomenon is known as auto-oscillation and can lead to very large flow rate and pressure fluctuations in the system. In boiler feed systems, discharge pressure oscillations with amplitudes as high as 14 *bar* have been reported informally. It is a genuinely dynamic instability in the sense described in the last section, for it occurs when the slope of the pump head rise/flow rate curve is still strongly negative. Another characteristic of auto-oscillation is that it appears to occur more readily when the inducer is more heavily loaded; in other words at lower flow coefficients. These are also the circumstances under which backflow will occur. Indeed, Badowski (1969) puts forward the hypothesis that the dynamics of the backflow are responsible for cavitating inducer instability. Further evidence of this connection is provided by Hartmann and Soltis (1960) but with an atypical inducer that has 19 blades. It is certainly the case that the limit cycle associated with a strong auto-oscillation appears to involve large periodic oscillations in the backflow. Consequently, it would seem that any nonlinear model purporting to predict the magnitude of auto-oscillation should incorporate the dynamics of the backflow. While most of the detailed investigations have focussed on axial pumps and inducers, Yamamoto (1991) has observed and investigated auto-oscillation occurring in cavitating centrifugal pumps. He also noted the important role played by the backflow in the dynamics of the auto-oscillation.

Unlike compressor surge, the frequency of auto-oscillation, Ω_A, usually scales with the shaft speed of the pump. Figure 8.8 demonstrates this by plotting Ω_A/Ω against the shaft rpm ($60\Omega/2\pi$) for a particular helical inducer. Figure 8.9 (also from Braisted and Brennen 1980) shows how this reduced auto-oscillation frequency, Ω_A/Ω, varies with flow coefficient, ϕ, cavitation number, σ, and impeller geometry. While still noting that the frequency, Ω_A, will, in general, be system dependent, nevertheless the expression

$$\Omega_A/\Omega = (2\sigma)^{\frac{1}{2}} \tag{8.8}$$

appears to provide a crude estimate of the auto-oscillation frequency.

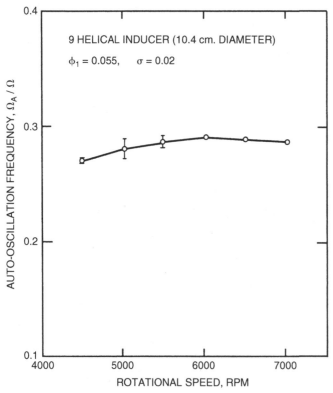

Figure 8.8. Data from Braisted and Brennen (1980) on the ratio of the auto-oscillation frequency to the shaft frequency as a function of the latter for a 9° helical inducer operating at a cavitation number of 0.02 and a flow coefficient of 0.055.

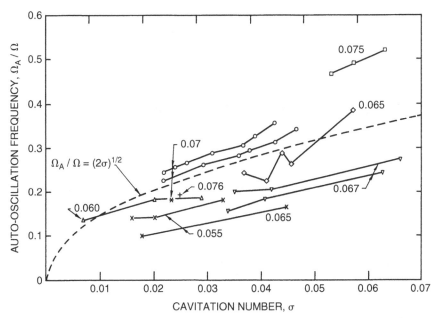

Figure 8.9. Data from Braisted and Brennen (1980) on the ratio of the frequency of auto-oscillation to the frequency of shaft rotation for several inducers: SSME Low Pressure LOX Pump models: 7.62 *cm* diameter: $\times$ (9000 *rpm*) and $+$ (12000 *rpm*), 10.2 *cm* diameter: $\circ$ (4000 *rpm*) and $\square$ (6000 *rpm*); 9° helical inducers: 7.58 *cm* diameter: $*$ (9000 *rpm*): 10.4 *cm* diameter: $\triangledown$ (with suction line flow straightener) and $\triangle$ (without suction line flow straightener). The flow coefficients, ϕ_1, are as labeled.

Some data from Yamamoto (1991) on the frequencies of auto-oscillation of a cavitating centrifugal pump are presented in figure 8.10. This data exhibits a dependence on the length of the suction pipe that reinforces the understanding of auto-oscillation as a system instability. The figure also shows the limits of instability observed by Yamamoto; these are unusual in that there appear to be two separate regions or zones of instability. Finally, it is clear that the data of figures 8.9 and 8.10 show a similar dependence of the auto-oscillation frequency on the cavitation number, σ, though the magnitudes of σ differ considerably. However, it is likely that the relative sizes of the cavities are similiar in the two cases, and therefore that the correlation between the auto-oscillation frequency and the relative cavity size might be closer than the correlation with cavitation number.

As previously stated, auto-oscillation occurs when the region of cavitation head loss is approached as the cavitation number is decreased. Figure 8.11 provides an example of the limits of auto-oscillation taken from the work of Braisted and Brennen (1980). However, since the onset is even more dependent than the auto-oscillation frequency on the detailed dynamic characteristics of the system, it would not even be wise to quote any approximate guideline for onset. Our current understanding is that the methodologies of chapter 9 are essential for any prediction of auto-oscillation.

It should be noted that chapter 9 describes linear perturbation models that can predict the limits of oscillation but not the amplitude of the oscillation once it occurs.

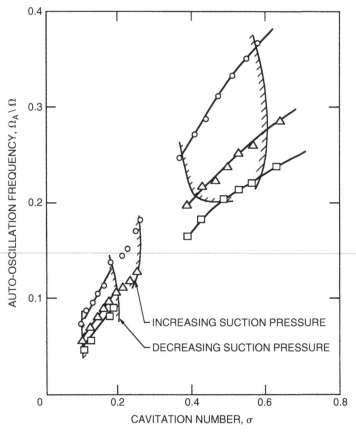

Figure 8.10. Data from Yamamoto (1991) on the ratio of the frequency of auto-oscillation to the frequency of shaft rotation for a centrifugal pump. Data is shown for three different lengths of suction pipe: ○ (2.7 *m*), △ (4.9 *m*), and □ (7.1 *m*). Regions of instability are indicated by the hatched lines.

There do not appear to be any accepted analytical models that can make this important prediction. Furthermore, the energy dissipated in the large amplitude oscillations within the pump can lead to a major change in the mean (time averaged) performance of the pump. One example of the effect of auto-oscillation on the head rise across a cavitating inducer is shown in figure 8.12 (from Braisted and Brennen 1980) which contains cavitation performance curves for three flow coefficients. The sequence of events leading to these results was as follows. For each flow rate, the cavitation number was decreased until the onset of auto-oscillation at the point labeled A, when the head immediately dropped to the point B (an unavoidable change in the pump inlet pressure and therefore in σ would often occur at the same time). Increasing the cavitation number again would not immediately eliminate the auto-oscillation. Instead the oscillations would persist until the cavitation number was raised to the value at the point C where the disappearance of auto-oscillation would cause recovery to the point D. This set of experiments demonstrate (i) that under auto-oscillation conditions (B, C) the head rise across this particular inducer was about half of the head rise

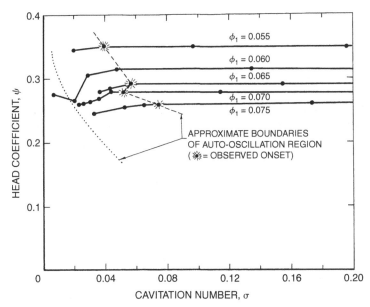

Figure 8.11. Cavitation performance of the SSME low pressure LOX pump model, Impeller IV, showing the onset and approximate desinence of the auto-oscillation at 6000 *rpm* (from Braisted and Brennen 1980).

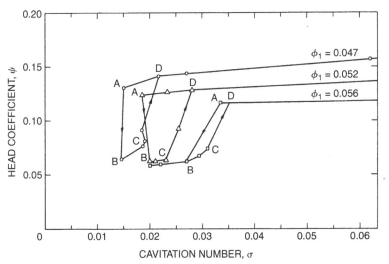

Figure 8.12. Data from a helical inducer illustrating the decrease in head with the onset of auto-oscillation ($A \rightarrow B$) and the auto-oscillation hysteresis occurring with subsequent increase in σ (from Braisted and Brennen 1980).

without auto-oscillation (A, D) and (ii) that a significant auto-oscillation hysteresis exists in which the auto-oscillation inception and desinence cavitation numbers can be significantly different. Neither of these nonlinear effects can be predicted by the frequency domain methods of chapter 9. In other inducers, the drop in head with the onset of auto-oscillation is not as large as in figure 8.12 but it is still present; it has also been reported by Rosenmann (1965). This effect may account for the somewhat jagged form of the cavitation characteristic as breakdown is approached.

8.8 Rotor-Stator Interaction: Flow Patterns

In section 8.2, we described the two basic frequencies of rotor-stator interaction: the excitation of the stator flow at $Z_R\Omega$ and the excitation of the rotor flow at $Z_S\Omega$. Apart from the superharmonics $mZ_R\Omega$ and $mZ_S\Omega$ that are generated by nonlinearities, subharmonics can also occur. When they do they can cause major problems, since the fluid and structural damping is smaller for these lower frequencies. To avoid such subharmonics, turbomachines are usually designed with blade numbers, Z_R and Z_S, which have small integer common factors.

The various harmonics of blade passage excitation can be visualized by generating an "encounter" (or interference) diagram that is a function only of the integers Z_R and Z_S. In these encounter diagrams, of which figures 8.13 and 8.14 are examples, each of the horizontal lines represents the position of a particular stator blade. The circular geometry has been unwrapped so that a passing rotor blade proceeds from top to bottom as it rotates past the stator blades. Each vertical line represents a moment in time, the period covered being one complete revolution of the rotor beginning at the far left and returning to that moment on the far right. Within this framework, the moment and position of all the rotor-stator blade encounters are shown by an "0." Such encounter diagrams allow one to examine the various frequencies and patterns generated by rotor-stator interactions and this is perhaps best illustrated by referring to the examples of figure 8.13 for the case of $Z_R = 6$, $Z_S = 7$, and figure 8.14 for the case of $Z_R = 6$, $Z_S = 16$. First, one can always follow the diagonal progress of individual rotor blades as indicated by lines such as those marked 1Ω in the examples. But other diagonal lines are also evident. For example, in figure 8.13 the perturbation consisting of a single cell, and propagating in the reverse direction at 6Ω is strongly indicated. Parenthetically we note that, in any machine in which $Z_S = Z_R + 1$, a

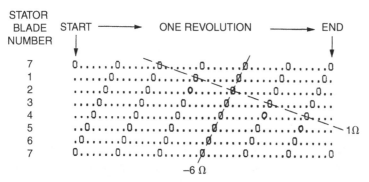

Figure 8.13. Encounter diagram for rotor-stator interaction in a turbomachine with $Z_R = 6$, $Z_S = 7$. Each row is for a specific stator blade and time runs horizontally covering one revolution as one proceeds from left to right. Encounters between a rotor blade and a stator blade are marked by an 0.

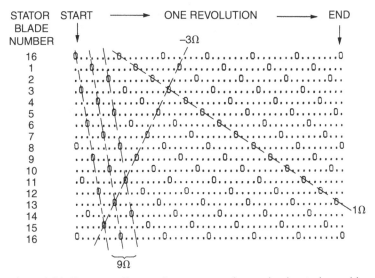

Figure 8.14. Encounter diagram for rotor-stator interaction in a turbomachine with $Z_R = 6$, $Z_S = 16$.

perturbation with a reverse speed of $-Z_R\Omega$ is always present. Also in figure 8.14, there are quite strong lines indicating an encounter pattern rotating at 9Ω and consisting of two diametrically opposite cells. Other propagating disturbance patterns are also suggested by figure 8.14. For example, the backward propagating disturbance rotating at 3Ω in the reverse direction and consisting of four equally spaced perturbation cells is indicated by the lines marked -3Ω. It is, of course, possible to connect up the encounter points in a very large number of ways, but clearly only those disturbances with a large number of encounters per cycle (high "density") will generate a large enough flow perturbation to be significant. However, among the top two or three possibilities, it is not necessarily a simple matter to determine which will manifest itself in the actual flow. That requires more detailed analysis of the flow.

The flow perturbations caused by blade passage excitation are nicely illustrated by Miyagawa *et al.* (1992) in their observations of the flows in high head pump turbines. One of the cases they explored was that of figure 8.14, namely $Z_R = 6$, $Z_S = 16$. Figure 8.15 has been extracted from the videotape of their unsteady flow observations and shows two diametrically opposite perturbation cells propagating around at nine times the impeller rotating speed, one of the "dense" perturbation patterns predicted by the encounter diagram of figure 8.14.

8.9 Rotor-Stator Interaction: Forces

When one rotor (or stator) blade passes through the wake of an upstream stator (or rotor) blade, it will clearly experience a fluctuation in the fluid forces that act upon it. In this section, the nature and magnitude of these rotor-stator interaction forces will be explored. Experience has shown that these unsteady forces are a strong function of

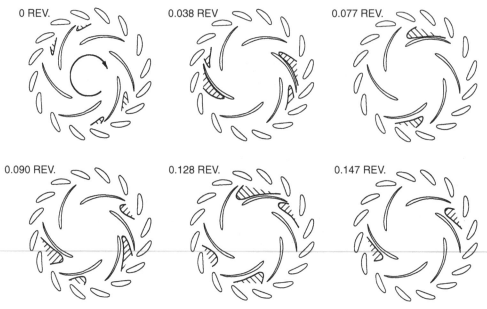

Figure 8.15. The propagation of a low pressure region (hatched) at nine times the impeller rotational speed in the flow through a high head pump-turbine. The sketches show six instants in time equally spaced within one sixth of a revolution. Made from videotape provided by Miyagawa *et al.* (1992).

the gap between the locus of the trailing edge of the upstream blade and the locus of the leading edge of the downstream blade. This distance will be termed the interblade spacing, and will be denoted by c_b.

Most axial compressors and turbines operate with fairly large interblade spacings, greater than 10% of the blade chord. As a result, the unsteady flows and forces measured under these circumstances (Gallus 1979, Gallus *et al.* 1980, Dring *et al.* 1982, Iino and Kasai 1985) are substantially smaller than those measured for radial machines (such as centrifugal pumps) in which the interblade spacing between the impeller and diffuser blades may be only a few percent of the impeller radius. Indeed, structural failure of the leading edge of centrifugal diffuser blades is not uncommon in the industry, and is typically solved by increasing the interblade spacing, though at the cost of reduced hydraulic performance.

Several early investigations of rotor-stator interaction forces were carried out using single foils in a wind tunnel (for example, Lefcort 1965). However, Gallus *et al.* (1980) appear to have been the first to measure the unsteady flows and forces due to rotor-stator interaction in an axial flow compressor. They attempt to collate their measurements with the theoretical analyses of Kemp and Sears (1955), Meyer (1958), Horlock (1968) and others. The measurements were conducted with large interblade spacing to axial chord ratios of about 50%, and involved documentation of the blade wakes. The impingement of these wakes on the following row of blades causes pressure fluctuations that are largest on the forward suction surface and small near the trailing

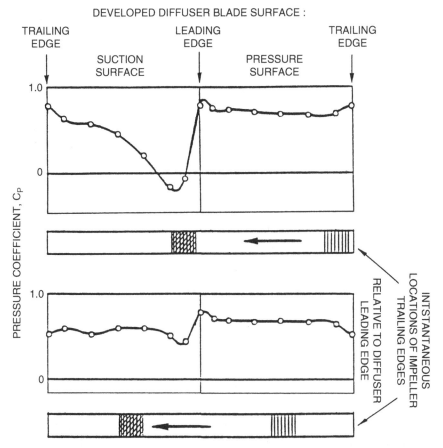

Figure 8.16. Pressure distributions on a diffuser blade at two different instants during the passage of an impeller blade. Data for an interblade spacing of 1.5% and $\phi_2 = 0.12$ (from Arndt *et al.* 1989).

edge of those blades. These pressure fluctuations lead to a fluctuation in the lift coefficient of ± 0.06. Moreover, Gallus *et al.* (1980) show that the forces vary roughly inversely with the interblade spacing to axial chord ratio. Extrapolation would suggest that the unsteady and steady components of the lift might be roughly the same if this ratio were decreased to 5%. This estimate is confirmed by the measurements of Arndt *et al.*, described below. Before concluding this discussion of rotor-stator interaction forces in axial flow machines, we note that Dring *et al.* (1982) have examined the flows and forces for an interblade spacing to axial chord ratio of 0.35 and obtained results similar to those of Gallus *et al.*

Recently, Arndt *et al.* (1989, 1990) (see also Brennen *et al.* 1988) have made measurements of the unsteady pressures and forces that occur in a radial flow machine when an impeller blade passes a diffuser blade. Figure 8.16 presents instantaneous pressure distributions (ensemble-averaged over many revolutions) for two particular relative positions of the impeller and diffuser blades. In the upper graph the trailing edge of the impeller blade has just passed the leading edge of the diffuser blade, causing

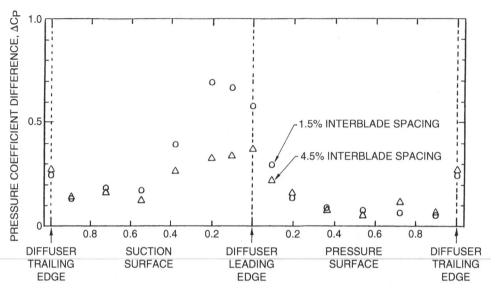

Figure 8.17. Magnitude of the fluctuation in the coefficient of pressure on a diffuser blade during the passage of an impeller blade as a function of location on the diffuser blade surface for two interblade spacings (from Arndt *et al.* 1989).

a large perturbation in the pressure on the suction surface of the diffuser blade. Indeed, in this example, the pressure over a small region has fallen below the impeller inlet pressure ($C_p < 0$). The lower graph is the pressure distribution at a later time when the impeller blade is about half-way to the next diffuser blade. The perturbation in the diffuser blade pressure distribution has largely dissipated. Closer examination of the data suggests that the perturbation takes the form of a wave of negative pressure traveling along the suction surface of the diffuser blade and being attenuated as it propagates. This and other observations suggest that the cause is a vortex shed from the leading edge of the diffuser blade by the passage of the trailing edge of the impeller. This vortex is then convected along the suction surface of the diffuser blade.

The difference between the maximum and minimum pressure coefficient, ΔC_p, experienced at each position on the surface of a diffuser blade is plotted as a function of position in figure 8.17. Data is shown for two interblade spacings, $c_b = 0.015 R_{T2}$ and $0.045 R_{T2}$. This figure reiterates the fact that the pressure perturbations are largest on the suction surface just downstream of the leading edge. It also demonstrates that the pressure perturbations for the 1.5% interblade spacing are about double those for the 4.5% interblade spacing. Figure 8.17 was obtained at a particular flow coefficient of $\phi_2 = 0.12$; however, the same phenomena were encountered in the range $0.05 < \phi_2 < 0.15$, and the magnitude of the pressure perturbation showed an increase of about 50% between $\phi_2 = 0.05$ and $\phi_2 = 0.15$.

Given both the magnitude and phase of the instantaneous pressures on the surface of a diffuser blade, the result may be integrated to obtain the instantaneous lift, L, on the diffuser blade. Here the lift coefficient is defined as $C_L = L/\frac{1}{2}\rho\Omega^2 R_{T2}^2 cb$ where

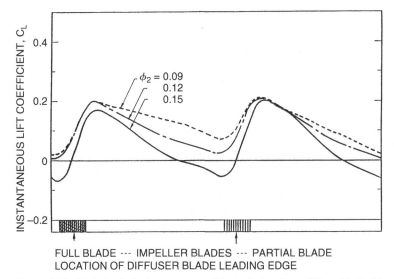

FULL BLADE --- IMPELLER BLADES --- PARTIAL BLADE
LOCATION OF DIFFUSER BLADE LEADING EDGE

Figure 8.18. Variation in the instantaneous lift coefficient for a diffuser blade. The position of the diffuser blade leading edge relative to the impeller blade trailing edge is also shown. The data is for an interblade spacing of 4.5% (from Arndt *et al.* 1989).

L is the force on the blade perpendicular to the mean chord, c is the chord, and b is the span of the diffuser blade. Time histories of C_L are plotted in figure 8.18 for three different flow coefficients and an interblade spacing of 4.5%. Since the impeller blades consisted of main blades separated by partial blades, two ensemble-averaged cycles are shown for C_L though the differences between the passage of a full blade and a partial blade are small. Notice that even for the larger 4.5% interblade spacing, the instantaneous lift can be as much as three times the mean lift. Consequently, a structural design criterion based on the mean lift on the blades would be seriously flawed. Indeed, in this case it is clear that the principal structural consideration should be the unsteady lift, not the steady lift.

Arndt *et al.* (1990) also examined the unsteady pressures on the upstream impeller blades for a variety of diffusers. Again, large pressure fluctuations were encountered as a result of rotor-stator interaction. Typical results are shown in figure 8.19 where the magnitude of the pressure fluctuations is presented as a function of flow coefficient for three different locations on the surface of an impeller blade: (i) on the flat of the trailing edge, (ii) on the suction surface at $r/R_{T2} = 0.937$, and (iii) on the pressure surface at $r/R_{T2} = 0.987$. The data are for a 5% interblade spacing and all data points represent ensemble averages. The magnitudes of the fluctuations are of the same order as the pressure fluctuations on the diffuser blades, indicating that the unsteady loads on the *upstream* blade in rotor-stator interaction can also be substantial. Note, however, that contrary to the trend with the diffuser blades, the magnitude of the pressure fluctuations *decrease* with increasing flow coefficient. Finally, note that the magnitude

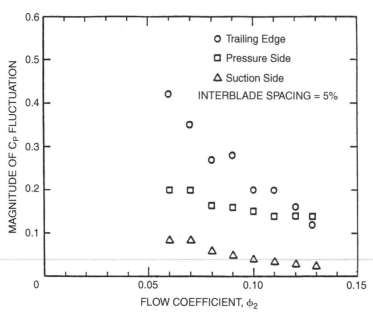

Figure 8.19. Magnitude of the fluctuations in the pressure coefficient at three locations near the trailing edge of an impeller blade during the passage of a diffuser blade (from Arndt *et al.* 1990).

of the pressure fluctuations are as large as the total head rise across the pump. This raises the possibility of transient cavitation being caused by rotor-stator interaction.

Considering the magnitude of these rotor-stator interaction effects, it is surprising that there is such a limited quantity of data available on the unsteady forces.

8.10 Developed Cavity Oscillation

There are several circumstances in which developed cavities can exhibit self-sustained oscillations in the absence of any external excitation. One of these is the instability associated with a partial cavity whose length is approximately equal to the chord of the foil. Experimentally, it is observed that when the cavitation number is decreased to the level at which the attached partial cavity on a single hydrofoil approaches about 0.7 of the chord, c, of the foil, the cavity will begin to oscillate violently (Wade and Acosta 1966). It will grow to a length of about $1.5c$, at which point the cavity will be pinched off at about $0.5c$, and the separated cloud will collapse as it is convected downstream. This collapsing cloud of bubbles carries with it shed vorticity, so that the lift on the foil oscillates at the same time. This phenomenon is called "partial cavitation oscillation." It persists with further decrease in cavitation number until a point is reached at which the cavity collapses at some critical distance downstream of the trailing edge that is usually about $0.3c$. For cavitation numbers lower than this, the flow again becomes quite stable. The frequency of partial cavitation oscillation on a single foil is usually less than $0.1U/c$, where U is the velocity of the oncoming stream, and c is the chord length of the foil. In cascades or pumps, supercavitation is usually only approached in machines of low solidity, but, under such circumstances, partial cavitation oscillation can occur, and can be quite violent. Wade and Acosta (1966) were the first to observe partial cavitation oscillation in a cascade. During another set of experiments on cavitating cascades, Young, Murphy, and Reddcliff (1972) observed only "random unsteadiness of the cavities."

One plausible explanation for this partial cavitation instability can be gleaned from the free streamline solutions for a cavitating foil that were described in section 7.8. The results from equations 7.9 to 7.12 can be used to plot the lift coefficient as a function of angle of attack for various cavitation numbers, as shown in figure 8.20. The results from both the partial cavitation and the supercavitation analyses are shown. Moreover, we have marked with a dotted line the locus of those points at which the supercavitating solution yields $dC_L/d\alpha = 0$; it is easily shown that this occurs when $\ell = 4c/3$. We have also marked with a dotted line the locus of those points at which the partial cavitation solution yields $dC_L/d\alpha = \infty$; it can also be shown that this occurs when $\ell = 3c/4$. Note that these dotted lines separate regions for which $dC_L/d\alpha > 0$ from that region in which $dC_L/d\alpha < 0$. Heuristically, it could be argued that $dC_L/d\alpha < 0$ implies an unstable flow. It would follow that the region between the dotted lines in

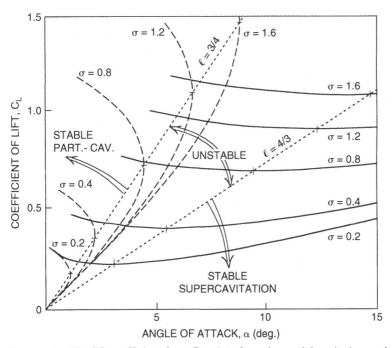

Figure 8.20. The lift coefficient for a flat plate from the partial cavitation analysis of Acosta (1955) (dashed lines) and the supercavitating analysis of Tulin (1953) (solid lines); C_L is shown as a function of angle of attack, α, for several cavitation numbers, σ. The dotted lines are the boundaries of the region in which the cavity length is between 3/4 and 4/3 of a chord, and in which $dC_L/d\alpha < 0$.

figure 8.20 represents a regime of unstable operation. The boundaries of this regime are $\frac{3}{4} < \frac{\ell}{c} < \frac{4}{3}$, and do, indeed, seem to correspond quite closely to the observed regime of unstable cavity oscillation (Wade and Acosta 1966).

A second circumstance in which a fully developed cavity may exhibit natural oscillations occurs when the cavity is formed by introducing air to the wake of a foil in order to form a "ventilated cavity." When the flow rate of air exceeds a certain critical level, the cavity may begin to oscillate, large pockets of air being shed at the rear of the main cavity during each cycle of oscillation. This problem was studied by Silberman and Song (1961) and by Song (1962). The typical radian frequency for these oscillations is about $6U/\ell$, based on the length of the cavity, ℓ. Clearly, this second phenomenon is less relevant to pump applications.

8.11 Acoustic Resonances

In the absence of cavitation or flow-induced vibration, flow noise generated within the turbomachine itself is almost never an issue when the fluid is a liquid. One reason for this is that the large wavelength of the sound in the liquid leads to internal acoustic resonances that are too high in frequency and, therefore, too highly damped to be important. This contrasts with the important role played by internal resonances in the production of noise in gas turbines and compressors (Tyler and Sofrin 1962, Cumpsty 1977). In noncavitating liquid turbomachinery, the higher acoustic velocity and the smaller acoustic damping mean that pipeline resonances play the same kind of role that the internal resonances play in the production of noise in gas turbomachinery.

In liquid turbomachines, resonances exterior to the machine or resonances associated with cavitation do create a number of serious vibration problems. As mentioned in the introduction, pipeline resonances with the acoustic modes of the inlet or discharge piping can occur when one of the excitation frequencies produced by the pump or hydraulic turbine happens to coincide with one of the acoustic modes of those pipelines. Jaeger (1963) and Strub (1963) document a number of cases of resonance in hydropower systems. Many of these do not involve excitation from the turbine but some do involve excitation at blade passing frequencies (Strub 1963). One of the striking features of these phenomena is that very high harmonics of the pipelines can be involved (20th harmonics have been noted) so that damage occurs at a whole series of nodes equally spaced along the pipeline. The cases described by Jaeger involve very large pressure oscillations, some of which led to major failures of the installation. Sparks and Wachel (1976) have similarly documented a number of cases of pipeline resonance in pumping systems. They correctly identify some of these as system instabilities of the kind discussed in section 8.6 and chapter 9.

Cavitation-induced resonances and vibration problems are dealt with in other sections of this chapter. But it is appropriate in the context of resonances to mention one other possible cavitation mechanism even though it has not, as yet, been demonstrated experimentally. One might judge that the natural frequency, ω_P, of bubbles given by equation 6.14 (section 6.5), being of the order of kHz, would be too high to cause vibration problems. However, it transpires that a finite cloud of bubbles may have much smaller natural frequencies that could resonant, for example, with a blade passage frequency to produce a problem. d'Agostino and Brennen (1983) showed that the lowest natural frequency, ω_C, of a spherical cloud of bubbles of radius, A, consisting of bubbles of radius, R, and with a void fraction of α would be given by

$$\omega_C = \omega_P \left[1 + \frac{4}{3\pi^2} \frac{A^2}{R^2} \frac{\alpha}{1-\alpha} \right]^{-\frac{1}{2}} \tag{8.9}$$

It follows that, if $\alpha A^2/R^2 \gg 1$, then the cloud frequency will be significantly smaller than the bubble frequency. This requires that the void fraction be sufficiently large so that $\alpha \gg R^2/A^2$. However, this could be relatively easily achieved in large clouds of small bubbles. Though the importance of cloud cavitation in pumps has been clearly demonstrated (see section 6.3), the role played by the basic dynamic characteristics of clouds has not, as yet, been elucidated.

8.12 Blade Flutter

Up to this point, all of the instabilities have been essentially hydrodynamic and would occur with a completely rigid structure. However, it needs to be observed that structural flexibility could modify any of the phenomena described. Furthermore, if a hydrodynamic instability frequency happens to coincide with the frequency of a major mode of vibration of the structure, the result will be a much more dangerous vibration problem. Though the hydroelastic behavior of single hydrofoils has been fairly well established (see the review by Abramson 1969), it would be virtually impossible to classify all of the possible fluid-structure interactions in a turbomachine given the number of possible hydrodynamic instabilities and the complexity of the typical pump structure. Rather, we shall confine attention to one of the simpler interactions and briefly discuss blade flutter. Though the rotor-stator interaction effects outlined above are more likely to cause serious blade vibration problems in turbomachines, it is also true that a blade may flutter and fail even in the absence of such excitation.

It is well known (see, for example, Fung 1955) that the incompressible, unstalled flow around a single airfoil will not exhibit flutter when permitted only one degree of freedom of flutter motion. Thus, classic aircraft wing flutter requires the coupling of two degrees of flutter motion, normally the bending and torsional modes of the cantilevered wing. Turbomachinery flutter is quite different from classic aircraft wing flutter and usually involves the excitation of a single structural mode. Several different phenomena can lead to single degree of freedom flutter when it would not otherwise occur in incompressible, unseparated (unstalled) flow. First, there are the effects of compressibility that can lead to phenomena such as supersonic flutter and choke flutter. These have been the subject of much research (see, for example, the reviews of Miko-lajczak *et al.* (1975), Platzer (1978), Sisto (1977), McCroskey (1977)), but are not of direct concern in the context of liquid turbomachinery, though the compressibility introduced by cavitation might provide some useful analogies. Of greater importance in the context of liquid turbomachinery is the phenomenon of stall flutter (see, for example, Sisto 1953, Fung 1955). A blade which is stalled during all or part of a cycle of oscillation can exhibit single degree of freedom flutter, and this type of flutter has been recognized as a problem in turbomachinery for many years (Platzer 1978, Sisto 1977). Unfortunately, there has been relatively little analytical work on stall flutter and any modern theory must at least consider the characteristics of dynamic stall (see McCroskey 1977). Like all single degree of freedom flutter problems, including those in turbomachines, the critical incident speed for the onset of stall flutter, U_C, is normally given by a particular value of a reduced speed, $U_{CR} = 2U_C/c\omega_F$, where c is the chord length and ω_F is the frequency of flutter or the natural frequency of the participating structural vibration mode. The inverse of U_{CR} is the reduced frequency, k_{CR}, or Strouhal number. Fung (1955) points out that the reduced frequency for stall flutter with a single foil is a function of the difference, θ, between the mean angle of incidence of the flow and the static angle of stall. A crude guide would be $k_{CR} = 0.3 + 4.5\theta$, $0.1 < k_{CR} < 0.8$. The second term in the expression for k_{CR} reflects the decrease in the critical speed with increasing incidence.

Of course, in a turbomachine or cascade, the vibration of one blade will generate forces on the neighboring blades (see, for example, Whitehead 1960), and these interactions can cause significant differences in the flutter analyses and critical speeds; often they have a large unfavorable effect on the flutter characteristics (McCroskey 1977). One must allow for various phase angles between neighboring blades, and examine waves which travel both forward and backward relative to the rotation of the rotor. A complete analysis of the vibrational modes of the rotor (or stator) must be combined with an unsteady fluid flow analysis (see, for example, Verdon 1985) in order to accurately predict the flutter boundaries in a turbomachine. Of course, most of the literature deals with structures that are typical of compressors and turbines. The lowest modes of vibration in a pump, on the other hand, can be very different in character from those in a compressor or turbine. Usually the blades have a much smaller aspect ratio so that the lowest modes involve localized vibration of the leading or trailing edges of the blades. Consequently, any potential flutter is likely to cause failure of portions of these leading or trailing edges.

The other major factor is the effect of cavitation. The changes which developed cavitation cause in the lift and drag characteristics of a single foil, also cause a fundamental change in the flutter characteristics with the result that a single cavitating foil can flutter (Abramson 1969). Thus a cavitating foil is unlike a noncavitating, nonseparated foil but qualitatively similar to a stalled foil whose flow it more closely resembles. Abramson (1969) provides a useful review of both the experiments and the

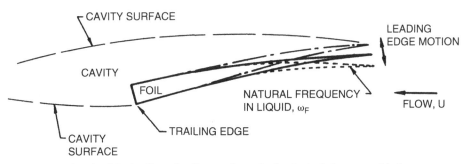

Figure 8.21. Sketch of the leading edge flutter of a cavitating hydrofoil or pump blade.

analyses of flutter of rigid cavitating foils. However, as we previously remarked, the most likely form of flutter in a pump will not involve global blade motion but flexure of the leading or trailing edges. Since cavitation occurs at the leading edges, and since these are often made thin in order to optimize the hydraulic performance, leading edge flutter seems the most likely concern (figure 8.21). Data on this phenomenon was obtained by Brennen, Oey, and Babcock (1980), and is presented in figure 8.22. The critical incident fluid velocity, U_C, is nondimensionalized using ω_F, the lowest natural frequency of oscillation of the leading edge immersed in water, and a dimension, c_F, that corresponds to the typical chordwise length of the foil from the leading edge to the first node of the first mode of vibration. The data shows that $U_C/c_F\omega_F$ is almost independent of the incidence angle, and is consistent for a wide range of natural frequencies. Brennen et $al.$ also utilize the unsteady lift and moment coefficients calculated by Parkin (1962) to generate a theoretical estimate of $U_C/c_F\omega_F$ of 0.14. From figure 8.22 this seems to constitute an upper design limit on the reduced critical speed. Also note that the value of 0.14 is much smaller than the values of 1–3 quoted earlier for the stall flutter of a noncavitating foil. This difference emphasizes the enhanced flutter possibilities caused by cavitation. Brennen et $al.$ also tested their foils under noncavitating conditions but found no sign of flutter even when the tunnel velocity was much larger than the cavitating flutter speed.

One footnote on the connection between the flutter characteristics of figure 8.22 and the partial cavitation oscillation of section 8.10 is worth adding. The data of figure 8.22 was obtained with long attached cavities, covering the entire suction surface of the foil as indicated in figure 8.21. At larger cavitation numbers, when the cavity length was decreased to about two chord lengths, the critical speed decreased markedly, and the leading edge flutter phenomenon began to metamorphose into the partial cavitation oscillation described in section 8.10.

8.13 Pogo Instabilities

All of the other discussion in this chapter has assumed that the turbomachine as a whole remains fixed in a nonaccelerating reference frame or, at least, that a vibrational degree

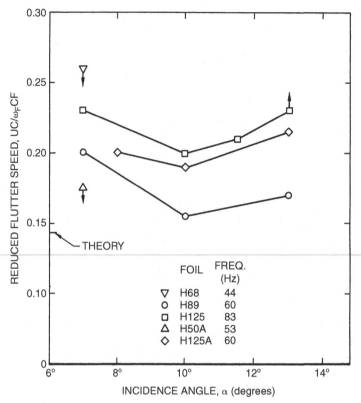

Figure 8.22. Dimensionless critical flutter speeds for single supercavitating hydrofoils at various angles of incidence, α, and very long cavities (> 5 chord lengths) (from Brennen, Oey, and Babcock 1980).

of freedom of the machine is not necessary for the instability to occur. However, when a mechanism exists by which the internal flow and pressure oscillations can lead to vibration of the turbomachine as a whole, then a new set of possibilities are created. We refer to circumstances in which flow or pressure oscillations lead to vibration of the turbomachine (or its inlet or discharge pipelines) which in turn generate pressure oscillations that feed back to create instability. An example is the class of liquid-propelled rocket vehicle instabilities known as Pogo instabilities (NASA 1970). Here the longitudinal vibration of the rocket causes flow and pressure oscillation in the fuel tanks and, therefore, in the inlet lines. This, in turn, implies that the engines experience fluctuating inlet conditions, and as a result they produce a fluctuating thrust that promotes the longitudinal vibration of the vehicle. Rubin (1966) and Vaage et al. (1972) provide many of the details of these phenomena that are beyond the scope of this text. It is, however, important to note that the dynamics of the cavitating inducer pumps are crucial in determining the limits of these Pogo instabilities, and provide one of the main motivations for the measurements of the dynamic transfer functions of cavitating inducers described in chapter 9.

In closing, it is important to note that feedback systems involving vibrational motion of the turbomachine are certainly not confined to liquid propelled rockets. However, detailed examinations of the instabilities are mostly confined to this context. In section 9.15 of the next chapter, we provide a brief introduction to the frequency domain methods which can be used to address problems involving oscillatory, translational or rotational motions of the whole hydraulic sytem.

9

Unsteady Flow in Hydraulic Systems

9.1 Introduction

This chapter is devoted to a description of the methods available for the analysis of unsteady flows in pumps and their associated hydraulic systems. There are two basic approaches to the solution of unsteady internal flows: solution in the time domain or in the frequency domain. The traditional time domain methods for hydraulic systems are treated in depth elsewhere (for example, Streeter and Wylie 1967, 1974), and will only be touched upon here. They have the great advantage that they can incorporate the nonlinear convective inertial terms in the equations of fluid flow, and are best suited to evaluating the transient response of flows in long pipes in which the equations of the flow and the structure are fairly well established. However, they encounter great difficulties when either the geometry is complex (for example inside a pump), or the fluid is complex (for example in the presence of cavitation). Under these circumstances, frequency domain methods have distinct advantages, both analytically and experimentally. On the other hand, the nonlinear convective inertial terms cannot readily be included in the frequency-domain methodology and, consequently, these methods are only accurate for small perturbations from the mean flow. This does permit evaluation of stability limits, but not the evaluation of the amplitude of large unstable motions.

It should be stressed that many unsteady hydraulic system problems can and should be treated by the traditional time domain or "water-hammer" methods. However, since the focus of this monograph is on pumps and cavitation, we place an emphasis here on frequency domain methods. Sections 9.5 through 9.10 constitute an introduction to these frequency domain methods. This is followed by a summary of the transfer functions for simple components and for pumps, both noncavitating and cavitating. Up to the beginning of section 9.15, it is assumed that the hydraulic system is at rest in some inertial or nonaccelerating frame. However, as indicated in section 8.13, there is an important class of problems in which the hydraulic system itself is oscillating in space. In section 9.15, we present a brief introduction to the treatment of this class of problems.

9.2 Time Domain Methods

The application of time domain methods to one-dimensional fluid flow normally consists of the following three components. First, one establishes conditions for the conservation of mass and momentum in the fluid. These may be differential equations (as in the example in the next section) or they may be jump conditions (as in the analysis of a shock). Second, one must establish appropriate thermodynamic constraints governing the changes of state of the fluid. In almost all practical cases of single-phase flow, it is appropriate to assume that these changes are adiabatic. However, in multiphase flows the constraints can be much more complicated. Third, one must determine the response of the containing structure to the pressure changes in the fluid.

The analysis is made a great deal simpler in those circumstances in which it is accurate to assume that both the fluid and the structure behave barotropically. By definition, this implies that the change of state of the fluid is such that some thermodynamic quantity (such as the entropy) remains constant, and therefore the fluid density, $\rho(p)$, is a simple algebraic function of just one thermodynamic variable, for example the pressure. In the case of the structure, the assumption is that it deforms quasistatically, so that, for example, the cross-sectional area of a pipe, $A(p)$, is a simple, algebraic function of the fluid pressure, p. Note that this neglects any inertial or damping effects in the structure.

The importance of the assumption of a barotropic fluid and structure lies in the fact that it allows the calculation of a single, unambiguous speed of sound for waves traveling through the piping system. The sonic speed in the fluid alone is given by c_∞ where

$$c_\infty = (d\rho/dp)^{-\frac{1}{2}} \tag{9.1}$$

In a liquid, this is usually calculated from the bulk modulus, $\kappa = \rho/(d\rho/dp)$, since

$$c_\infty = (\kappa/\rho)^{-\frac{1}{2}} \tag{9.2}$$

However the sonic speed, c, for one-dimensional waves in a fluid-filled duct is influenced by the compressibility of both the liquid and the structure:

$$c = \pm\left[\frac{1}{A}\frac{d(\rho A)}{dp}\right]^{-\frac{1}{2}} \tag{9.3}$$

or, alternatively,

$$\frac{1}{\rho c^2} = \frac{1}{\rho c_\infty^2} + \frac{1}{A}\left(\frac{dA}{dp}\right) \tag{9.4}$$

The left-hand side is the acoustic impedance of the system, and the equation reveals that this is the sum of the acoustic impedance of the fluid alone, $1/\rho c_\infty^2$, plus an "acoustic impedance" of the structure given by $(dA/dp)/A$. For example, for a thin-walled pipe made of an elastic material of Young's modulus, E, the acoustic impedance of the structure is $2a/E\delta$, where a and δ are the radius and the wall thickness of the pipe ($\delta \ll a$). The resulting form of equation 9.4,

$$c = \left[\frac{1}{c_\infty^2} + \frac{2\rho a}{E\delta}\right]^{-\frac{1}{2}} \tag{9.5}$$

is known as the Joukowsky water hammer equation. It leads, for example, to values of c of about $1000\ m/s$ for water in standard steel pipes compared with $c_\infty \approx 1400\ m/s$. Other common expressions for c are those used for thick-walled tubes, for concrete tunnels, or for reinforced concrete pipes (Streeter and Wylie 1967).

9.3 Wave Propagation in Ducts

In order to solve unsteady flows in ducts, an expression for the sonic speed is combined with the differential form of the equation for conservation of mass (the continuity equation)

$$\frac{\partial}{\partial t}(\rho A) + \frac{\partial}{\partial s}(\rho A u) = 0 \tag{9.6}$$

where $u(s,t)$ is the cross-sectionally averaged or volumetric velocity, s is a coordinate measured along the duct, and t is time. The appropriate differential form of the momentum equation is

$$\rho\left[\frac{\partial u}{\partial t} + u\frac{\partial u}{\partial s}\right] = -\frac{\partial p}{\partial s} - \rho g_s - \frac{\rho f u|u|}{4a} \tag{9.7}$$

where g_s is the component of the acceleration due to gravity in the s direction, f is the friction factor, and a is the radius of the duct.

Now the barotropic assumption 9.3 allows the terms in equation 9.6 to be written as

$$\frac{\partial}{\partial t}(\rho A) = \frac{A}{c^2}\frac{\partial p}{\partial t}; \qquad \frac{\partial(\rho A)}{\partial s} = \frac{A}{c^2}\frac{\partial p}{\partial s} + \rho\frac{\partial A}{\partial s}\Big|_p \tag{9.8}$$

so the continuity equation becomes

$$\frac{1}{c^2}\frac{\partial p}{\partial t} + \frac{u}{c^2}\frac{\partial p}{\partial s} + \rho\left[\frac{\partial u}{\partial s} + \frac{u}{A}\frac{\partial A}{\partial s}\Big|_p\right] = 0 \tag{9.9}$$

Equations 9.7 and 9.9 are two simultaneous, first order, differential equations for the two unknown functions, $p(s,t)$ and $u(s,t)$. They can be solved given the barotropic relation for the fluid, $\rho(p)$, the friction factor, f, the normal cross-sectional area of the pipe, $A_0(s)$, and boundary conditions which will be discussed later. Normally the last term in equation 9.9 can be approximated by $\rho u (dA_0/ds)/A_0$. Note that c may be a function of s.

In the time domain methodology, equations 9.7 and 9.9 are normally solved using the method of characteristics (see, for example, Abbott 1966). This involves finding moving coordinate systems in which the equations may be written as ordinary rather than partial differential equations. Consider the relation that results when we multiply equation 9.9 by λ and add it to equation 9.7:

$$\rho \left[\frac{\partial u}{\partial t} + (u + \lambda) \frac{\partial u}{\partial s} \right] + \frac{\lambda}{c^2} \left[\frac{\partial p}{\partial t} + \left(u + \frac{c^2}{\lambda} \right) \frac{\partial p}{\partial s} \right]$$

$$+ \frac{\rho u \lambda}{A_0} \frac{dA_0}{ds} + \rho g_s + \frac{\rho f |u| u}{4a} = 0 \tag{9.10}$$

If the coefficients of $\frac{\partial u}{\partial s}$ and $\frac{\partial p}{\partial s}$ inside the square brackets were identical, in other words if $\lambda = \pm c$, then the expressions in the square brackets could be written as

$$\frac{\partial u}{\partial t} + (u \pm c) \frac{\partial u}{\partial s} \quad \text{and} \quad \frac{\partial p}{\partial t} + (u \pm c) \frac{\partial p}{\partial s} \tag{9.11}$$

and these are the derivatives $\frac{du}{dt}$ and $\frac{dp}{dt}$ on $\frac{ds}{dt} = u \pm c$. These lines $\frac{ds}{dt} = u \pm c$ are the characteristics, and on them we may write:

1. In a frame of reference moving with velocity $u + c$ or on $\frac{ds}{dt} = u + c$:

$$\frac{du}{dt} + \frac{1}{\rho c} \frac{dp}{dt} + \frac{uc}{A_0} \frac{dA_0}{ds} + g_s + \frac{f u |u|}{4a} = 0 \tag{9.12}$$

2. In a frame of reference moving with velocity $u - c$ or on $\frac{ds}{dt} = u - c$:

$$\frac{du}{dt} - \frac{1}{\rho c} \frac{dp}{dt} - \frac{uc}{A_0} \frac{dA_0}{ds} + g_s + \frac{f u |u|}{4a} = 0 \tag{9.13}$$

A simpler set of equations result if the piezometric head, h^*, defined as

$$h^* = \frac{p}{\rho g} + \int \frac{g_s}{g} ds \tag{9.14}$$

is used instead of the pressure, p, in equations 9.12 and 9.13. In almost all hydraulic problems of practical interest $p/\rho_L c^2 \ll 1$ and, therefore, the term $\rho^{-1} dp/dt$ in equations 9.12 and 9.13 may be approximated by $d(p/\rho)/dt$. It follows that on the two characteristics

$$\frac{1}{\rho c} \frac{dp}{dt} \pm g s \approx \frac{g}{c} \frac{dh^*}{dt} - \frac{u}{c} g_s \tag{9.15}$$

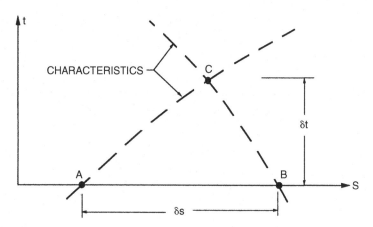

Figure 9.1. Method of characteristics.

and equations 9.12 and 9.13 become

1. On $\frac{ds}{dt} = u + c$

$$\frac{du}{dt} + \frac{g}{c}\frac{dh^*}{dt} + uc\frac{1}{A_0}\frac{dA_0}{ds} - \frac{ug_s}{c} + \frac{f}{4a}u|u| = 0 \tag{9.16}$$

2. On $\frac{ds}{dt} = u - c$

$$\frac{du}{dt} - \frac{g}{c}\frac{dh^*}{dt} - uc\frac{1}{A_0}\frac{dA_0}{ds} + \frac{ug_s}{c} + \frac{f}{4a}u|u| = 0 \tag{9.17}$$

These are the forms of the equations conventionally used in unsteady hydraulic water-hammer problems (Streeter and Wylie 1967). They are typically solved by relating the values at a time $t + \delta t$ (for example, point C of figure 9.1) to known values at the points A and B at time t. The lines AC and BC are characteristics, so the following finite difference forms of equations 9.16 and 9.17 apply:

$$\frac{(u_C - u_A)}{\delta t} + \frac{g}{c_A}\frac{(h_C^* - h_A^*)}{\delta t} + u_A c_A \left(\frac{1}{A_o}\frac{dA_0}{ds}\right)_A - \frac{u_A (g_s)_A}{c_A} + \frac{f_A u_A |u_A|}{4a} = 0 \tag{9.18}$$

and

$$\frac{(u_C - u_B)}{\delta t} - \frac{g}{c_B}\frac{(h_C^* - h_B^*)}{\delta t} - u_B c_B \left(\frac{1}{A_o}\frac{dA_0}{ds}\right)_B + \frac{u_B (g_s)_B}{c_B} + \frac{f_B u_B |u_B|}{4a} = 0 \tag{9.19}$$

If $c_A = c_B = c$, and the pipe is uniform, so that $dA_0/ds = 0$ and $f_A = f_B = f$, then these reduce to the following expressions for u_C and h_C^*:

$$u_C = \frac{(u_A + u_B)}{2} + \frac{g}{2c}\left(h_A^* - h_B^*\right) + \frac{\delta t}{2c}\left[u_A(g_s)_A - u_B(g_s)_B\right]$$
$$- \frac{f\delta t}{8a}\left[u_A|u_A| + u_B|u_B|\right] \tag{9.20}$$

$$h_C^* = \frac{\left(h_A^* + h_B^*\right)}{2} + \frac{c}{2g}\left(u_A - u_B\right) + \frac{\delta t}{2g}\left[u_A(g_s)_A + u_B(g_s)_B\right]$$
$$- \frac{fc\delta t}{8ag}\left[u_A|u_A| - u_B|u_B|\right] \tag{9.21}$$

9.4 Method of Characteristics

The typical numerical solution by the method of characteristics is depicted graphically in figure 9.2. The time interval, δt, and the spatial increment, δs, are specified. Then, given all values of the two dependent variables (say u and h^*) at one instant in time, one proceeds as follows to find all the values at points such as C at a time δt later. The intersection points, A and B, of the characteristics through C are first determined. Then interpolation between the known values at points such as R, S and T are used to determine the values of the dependent variables at A and B. The values at C follow from equations such as 9.20 and 9.21 or some alternative version. Repeating this for all points at time $t + \delta t$ allows one to march forward in time.

There is, however, a maximum time interval, δt, that will lead to a stable numerical solution. Typically this requires that δt be less than $\delta x/c$. In other words, it requires that the points A and B of figure 9.2 lie *inside* of the interval RST. The reason for this condition can be demonstrated in the following way. Assume for the sake of simplicity

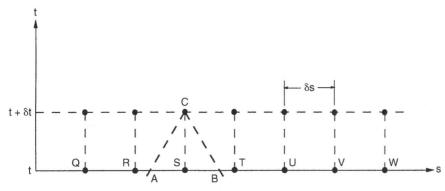

Figure 9.2. Example of numerical solution by method of characteristics.

that the slopes of the characteristics are $\pm c$; then the distances $AS = SB = c\delta t$. Using linear interpolation to find u_A and u_B from u_R, u_S and u_T leads to

$$u_A = u_S \left(1 - c\frac{\delta t}{\delta s}\right) + u_R \frac{c\delta t}{\delta s}$$

$$u_B = u_S \left(1 - c\frac{\delta t}{\delta s}\right) + u_T \frac{c\delta t}{\delta s} \qquad (9.22)$$

Consequently, an error in u_R of, say, δu would lead to an error in u_A of $c\delta u\delta t/\delta s$ (and similarly for u_T and u_B). Thus the error would be magnified with each time step unless $c\delta t/\delta s < 1$ and, therefore, the numerical integration is only stable if $\delta t < \delta x/c$. In many hydraulic system analyses this places a quite severe restriction on the time interval δt, and often necessitates a large number of time steps.

A procedure like the above will also require boundary conditions to be specified at any mesh point which lies either, at the end of a pipe or, at a junction of the pipe with a pipe of different size (or a pump or any other component). If the points S and C in figure 9.2 were end points, then only one characteristic would lie within the pipe and only one relation, 9.18 or 9.19, can be used. Therefore, the boundary condition must provide a second relation involving u_C or h_C^* (or both). An example is an open-ended pipe for which the pressure and, therefore, h^* is known. Alternatively, at a junction between two sizes of pipe, the two required relations will come from one characteristic in each of the two pipes, plus a continuity equation at the junction ensuring that the values of uA_0 in both pipes are the same at the junction. For this reason it is sometimes convenient to rewrite equations 9.16 and 9.17 in terms of the volume flow rate $Q = uA_0$ instead of u so that

1. On $\frac{ds}{dt} = u + c$

$$\frac{dQ}{dt} + \frac{A_0 g}{c}\frac{dh^*}{dt} + \frac{Qc}{A_0}\frac{dA_0}{ds} - \frac{Qg_s}{c} + \frac{f}{4aA_0}Q|Q| = 0 \qquad (9.23)$$

2. On $\frac{ds}{dt} = u - c$

$$\frac{dQ}{dt} - \frac{A_0 g}{c}\frac{dh^*}{dt} - \frac{Qc}{A_0}\frac{dA_0}{ds} + \frac{Qg_s}{c} + \frac{f}{4aA_0}Q|Q| = 0 \qquad (9.24)$$

Even in simple pipe flow, additional complications arise when the instantaneous pressure falls below vapor pressure and cavitation occurs. In the context of water-hammer analysis, this is known as "water column separation," and is of particular concern because the violent collapse of the cavity can cause severe structural damage (see, for example, Martin 1978). Furthermore, the occurrence of water column separation can trigger a series of cavity formations and collapses, resulting in a series of impulsive loads on the structure. The possibility of water column separation can be tracked by following the instantaneous pressure. To proceed beyond this point requires a procedure to incorporate a cavity in the waterhammer calculation using the method of characteristics. A number of authors (for example, Tanahashi and Kasahara 1969, Weyler *et al.* 1971, Safwat and van der Polder 1973) have shown that this is possible. However the calculated results after the first collapse can deviate substantially from the observations. This is probably due to the fact that the first cavity is often a single, coherent void. This will shatter into a cloud of smaller bubbles as a result of the violence of the first collapse. Subsequently, one is dealing with a bubbly medium whose wave propagation speeds may differ significantly from the acoustic speed assumed in the analytical model. Other studies have shown that qualitatively similar changes in the water-hammer behavior occur when gas bubbles form in the liquid as a result of dissolved gas coming out of solution (see, for example, Wiggert and Sundquist 1979).

In many time domain analyses, turbomachines are treated by assuming that the temporal rates of change are sufficiently slow that the turbomachine responds quasistatically, moving from one steady state operating point to another. Consequently, if points A and B lie at inlet to and discharge from the turbomachine then the equations relating the values at A and B would be

$$Q_B = Q_A = Q \qquad (9.25)$$

$$h_B^* = h_A^* + H(Q) \qquad (9.26)$$

where $H(Q)$ is the head rise across the machine at the flow rate, Q. Data presented later will show that the quasistatic assumption is only valid for rates of change less than about one-tenth the frequency of shaft rotation. For frequencies greater than this, the pump dynamics become important (see section 9.13).

For more detailed accounts of the methods of characteristics the reader is referred to Streeter and Wylie (1967), or any modern text on numerical methods. Furthermore, there are a number of standard codes available for time domain analysis of transients in hydraulic systems, such as that developed by Amies, Levek and Struesseld (1977). The methods work well so long as one has confidence in the differential equations and models which are used. In other circumstances, such as occur in two-phase flow, in cavitating flow, or in the complicated geometry of a turbomachine, the time domain methods may be less useful than the alternative frequency domain methods to which we now turn.

9.5 Frequency Domain Methods

When the quasistatic assumption for a device like a pump or turbine becomes questionable, or when the complexity of the fluid or the geometry makes the construction of a set of differential equations impractical or uncertain, then it is clear that experimental information on the dynamic behavior of the device is necessary. In practice, such experimental information is most readily obtained by subjecting the device to fluctuations in the flow rate or head for a range of frequencies, and measuring the fluctuating quantities at inlet and discharge. Such experimental results will be presented later. For present purposes it is sufficient to recognize that one practical advantage of frequency domain methods is the capability of incorporation of experimentally obtained dynamic information and the greater simplicity of the experiments required to obtain the necessary dynamic data. Another advantage, of course, is the core of fundamental knowledge that exists regarding such methodology (see, for example, Pipes 1940, Hennyey 1962, Paynter 1961, Brown 1967). As stated earlier, the disadvantage is that the methods are limited to small linear perturbations in the flow rate. When the perturbations are linear, Fourier analysis and synthesis can be used to convert from transient data to individual frequency components and vice versa. All the dependent variables such as the mean velocity, u, mass flow rate, m, pressure, p, or total pressure, p^T, are expressed as the sum of a mean component (denoted by an overbar) and a complex fluctuating component (denoted by a tilde) at a frequency, ω, which incorporates the amplitude and phase of the fluctuation:

$$p(s,t) = \bar{p}(s) + Re\left\{\tilde{p}(s,\omega)e^{j\omega t}\right\} \tag{9.27}$$

$$p^T(s,t) = \bar{p}^T(s) + Re\left\{\tilde{p}^T(s,\omega)e^{j\omega t}\right\} \tag{9.28}$$

$$m(s,t) = \bar{m}(s) + Re\left\{\tilde{m}(s,\omega)e^{j\omega t}\right\} \tag{9.29}$$

where j is $(-1)^{\frac{1}{2}}$ and Re denotes the real part. Since the perturbations are assumed linear ($|\tilde{u}| \ll \bar{u}, |\tilde{m}| \ll \bar{m}$, etc.), they can be readily superimposed, so a summation over many frequencies is implied in the above expressions. In general, the perturbation quantities will be functions of the mean flow characteristics as well as position, s, and frequency, ω.

We should note that there do exist a number of codes designed to examine the frequency response of hydraulic systems using frequency domain methods (see, for example, Amies and Greene 1977).

9.6 Order of the System

The first step in any unsteady flow analysis is to subdivide the system into components; the points separating two (or more) components will be referred to as system nodes. Typically, there would be nodes at the inlet and discharge flanges of a pump. Having done this, it is necessary to determine the order of the system, N, and this can be accomplished in one of several equivalent ways. The order of the system is the minimum number of independent fluctuating quantities which must be specified at a system node in order to provide a complete description of the unsteady flow at that location. It is also equal to the minimum number of independent, simultaneous first order differential equations needed to describe the fluid motion in, say, a length of pipe. In this summary we shall confine most of our discussion to systems of order two in which the dependent variables are the mass flow rate and either the pressure or the total head. This includes most of the common analyses of hydraulic systems. It is, however, important to recognize that order two systems are confined to

1. Incompressible flows at the system nodes, definable by pressure (or head), and flow rate.
2. Barotropic compressible flows in which, $\rho(p)$, so only the pressure (or head) and flow rate need be specified at system nodes. This category also includes those flexible structures for water-hammer analysis in which the local area is a function only of the local pressure. If, on the other hand, the local area depends on the area *and* the pressure elsewhere, then the system is of order 3 or higher.
3. Two-phase flows at the system nodes that can be represented by a *homogeneous flow model* that neglects the relative velocity between the phases. Any of the more accurate models that allow relative motion produce higher order systems.

Note that the order of the system can depend on the choice of system nodes. Consequently, an ideal evaporator or a condenser can be incorporated in an order two system provided the flow at the inlet node is single-phase (of type 2) and the flow at the discharge node also single-phase. A cavitating pump or turbine also falls within this category, provided the flow at both the inlet and discharge is pure liquid.

9.7 Transfer Matrices

The transfer matrix for any component or device is the matrix which relates the fluctuating quantities at the discharge node to the fluctuating quantities at the inlet node. The earliest exploration of such a concept in electrical networks appears to be due to Strecker and Feldtkeller (1929) while the utilization of the idea in the context of fluid systems owes much to the pioneering work of Pipes (1940). The concept is the following. If the quantities at inlet and discharge are denoted by subscripts $i = 1$ and $i = 2$, respectively, and, if $\{\tilde{q}_i^n\}, n = 1, 2 \to N$ denotes the vector of independent fluctuating quantities at inlet and discharge for a system of order N, then the transfer matrix, $[T]$, is defined as

$$\{\tilde{q}_2^n\} = [T]\{\tilde{q}_1^n\} \tag{9.30}$$

It is a square matrix of order N. For example, for an order two system in which the independent fluctuating variables are chosen to be the total pressure, $\tilde{p}^T$, and the mass flow rate, $\tilde{m}$, then a convenient transfer matrix is

$$\begin{Bmatrix} \tilde{p}_2^T \\ \tilde{m}_2 \end{Bmatrix} = \begin{bmatrix} T_{11} & T_{12} \\ T_{21} & T_{22} \end{bmatrix} \begin{Bmatrix} \tilde{p}_1^T \\ \tilde{m}_1 \end{Bmatrix} \tag{9.31}$$

The words transfer function and transfer matrix are used interchangeably here to refer to the matrix $[T]$. In general it will be a function of the frequency, ω, of the perturbations and the mean flow conditions in the device.

The most convenient independent fluctuating quantities for a hydraulic system of order two are usually

1. Either the pressure, $\tilde{p}$, or the instantaneous total pressure, $\tilde{p}^T$. Note that these are related by

$$\tilde{p}^T = \tilde{p} + \frac{\bar{u}^2}{2}\tilde{\rho} + \bar{\rho}\bar{u}\tilde{u} + gz\tilde{\rho} \qquad (9.32)$$

where $\bar{\rho}$ is the mean density, $\tilde{\rho}$ is the fluctuating density which is barotropically connected to $\tilde{p}$, and z is the vertical elevation of the system node. Neglecting the $\tilde{\rho}$ terms as is acceptable for incompressible flows

$$\tilde{p}^T = \tilde{p} + \bar{\rho}\bar{u}\tilde{u} \qquad (9.33)$$

2. Either the velocity, $\tilde{u}$, the volume flow rate, $\bar{A}\tilde{u} + \bar{u}\tilde{A}$, or the mass flow rate, $\tilde{m} = \bar{\rho}\bar{A}\tilde{u} + \bar{\rho}\bar{u}\tilde{A} + \bar{u}\bar{A}\tilde{\rho}$. Incompressible flow at a system node in a rigid pipe implies

$$\tilde{m} = \bar{\rho}\bar{A}\tilde{u} \qquad (9.34)$$

The most convenient choices are $\{\tilde{p},\tilde{m}\}$ or $\{\tilde{p}^T,\tilde{m}\}$, and, for these two vectors, we will respectively use transfer matrices denoted by $[T^*]$ and $[T]$, defined as

$$\begin{Bmatrix}\tilde{p}_2\\\tilde{m}_2\end{Bmatrix} = [T^*]\begin{Bmatrix}\tilde{p}_1\\\tilde{m}_1\end{Bmatrix}; \quad \begin{Bmatrix}\tilde{p}_2^T\\\tilde{m}_2\end{Bmatrix} = [T]\begin{Bmatrix}\tilde{p}_1^T\\\tilde{m}_1\end{Bmatrix} \qquad (9.35)$$

If the flow is incompressible and the cross-section at the nodes is rigid, then the $[T^*]$ and $[T]$ matrices are clearly connected by

$$T_{11} = T_{11}^* + \frac{\bar{u}_2}{A_2}T_{21}^*; \quad T_{12} = T_{12}^* - \frac{\bar{u}_1}{A_1}T_{11}^* + \frac{\bar{u}_2}{A_2}T_{22}^* - \frac{\bar{u}_1}{A_1}\frac{\bar{u}_2}{A_2}T_{21}^*$$

$$T_{21} = T_{21}^*; \quad T_{22} = T_{22}^* - \frac{\bar{u}_1}{A_1}T_{21}^* \qquad (9.36)$$

and hence one is readily constructed from the other. Note that the determinants of the two matrices, $[T]$ and $[T^*]$, are identical.

9.8 Distributed Systems

In the case of a distributed system such as a pipe, it is also appropriate to define a matrix $[F]$ (see Brown 1967) so that

$$\frac{d}{ds}\{\tilde{q}^n\} = -[F(s)]\{\tilde{q}^n\} \qquad (9.37)$$

Note that, apart from the frictional term, the equations 9.12 and 9.13 for flow in a pipe will lead to perturbation equations of this form. Furthermore, in many cases the frictional term is small, and can be approximated by a linear term in the perturbation equations; under such circumstances the frictional term will also fit into the form given by equation 9.37.

When the matrix $[F]$ is independent of location, s, the distributed system is called a "uniform system" (see section 9.10). For example, in equations 9.12 and 9.13, this would require ρ, c, a, f and A_0 to be approximated as constants (in addition to the linearization of the frictional term). Under such circumstances, equation 9.37 can be integrated over a finite length, ℓ, and the transfer matrix $[T]$ of the form 9.35 becomes

$$[T] = e^{-[F]\ell} \qquad (9.38)$$

where $e^{[F]\ell}$ is known as the "transmission matrix." For a system of order two, the explicit relation between $[T]$ and $[F]$ is

$$T_{11} = jF_{11}\left(e^{-j\lambda_2\ell} - e^{-j\lambda_1\ell}\right)/(\lambda_2 - \lambda_1)$$
$$+ \left(\lambda_2 e^{-j\lambda_1\ell} - \lambda_1 e^{-j\lambda_2\ell}\right)/(\lambda_2 - \lambda_1)$$
$$T_{12} = jF_{12}\left(e^{-j\lambda_2\ell} - e^{-j\lambda_1\ell}\right)/(\lambda_2 - \lambda_1)$$
$$T_{21} = jF_{21}\left(e^{-j\lambda_2\ell} - e^{-j\lambda_1\ell}\right)/(\lambda_2 - \lambda_1)$$
$$T_{22} = jF_{22}\left(e^{-j\lambda_2\ell} - e^{-j\lambda_1\ell}\right)/(\lambda_2 - \lambda_1)$$
$$+ \left(\lambda_2 e^{-j\lambda_2\ell} - \lambda_1 e^{-j\lambda_1\ell}\right)/(\lambda_2 - \lambda_1) \qquad (9.39)$$

where λ_1, λ_2 are the solutions of the equation

$$\lambda^2 + j\lambda(F_{11} + F_{22}) - (F_{11}F_{22} - F_{12}F_{21}) = 0 \qquad (9.40)$$

Some special features and properties of these transfer functions will be explored in the sections which follow.

9.9 Combinations of Transfer Matrices

When components are connected in series, the transfer matrix for the combination is clearly obtained by multiplying the transfer matrices of the individual components in the reverse order in which the flow passes through them. Thus, for example, the combination of a pump with a transfer matrix, $[TA]$, followed by a discharge line with a transfer matrix, $[TB]$, would have a system transfer matrix, $[TS]$, given by

$$[TS] = [TB][TA] \tag{9.41}$$

The parallel combination of two components is more complicated and does not produce such a simple result. Issues arise concerning the relations between the pressures of the inlet streams and the relations between the pressures of the discharge streams. Often it is appropriate to assume that the branching which creates the two inlet streams results in identical fluctuating total pressures at inlet to the two components, $\tilde{p}_1^T$. If, in addition, mixing losses at the downstream junction are neglected, so that the fluctuating total pressure, $\tilde{p}_2^T$, can be equated with the fluctuating total pressure at discharge from the two components, then the transfer function, $[TS]$, for the combination of two components (order two transfer functions denoted by $[TA]$ and $[TB]$) becomes

$$TS_{11} = (TA_{11}TB_{12} + TB_{11}TA_{12})/(TA_{12} + TB_{12})$$

$$TS_{12} = TA_{12}TB_{12}/(TA_{12} + TB_{12})$$

$$TS_{21} = TA_{21} + TB_{21}$$

$$\quad\quad - (TA_{11} - TB_{11})(TA_{22} - TB_{22})/(TA_{12} + TB_{12})$$

$$TS_{22} = (TA_{22}TB_{12} + TB_{22}TA_{12})/(TA_{12} + TB_{12}) \tag{9.42}$$

On the other hand, the circumstances at the junction of the two discharge streams may be such that the fluctuating static pressures (rather than the fluctuating total pressures) are equal. Then, if the inlet static pressures are also equal, the combined transfer matrix, $[TS^*]$, is related to those of the two components ($[TA^*]$ and $[TB^*]$) by the same relations as given in equations 9.42. Other combinations of choices are possible, but will not be detailed here.

Using the above combination rules, as well as the relations 9.36 between the $[T]$ and $[T*]$ matrices, the transfer functions for very complicated hydraulic networks can be systematically synthesized.

9.10 Properties of Transfer Matrices

Transfer matrices (and transmission matrices) have some fundamental properties that are valuable to recall when constructing or evaluating the dynamic properties of a component or system.

We first identify a "uniform" distributed component as one in which the differential equations (for example, equations 9.12 and 9.13 or 9.37) governing the fluid motion have coefficients which are independent of position, s. Then, for the class of systems represented by the equation 9.37, the matrix $[F]$ is independent of s. For a system of order two, the transfer function $[T]$ would take the explicit form given by equations 9.39.

To determine another property of this class of dynamic systems, consider that the equations 9.37 have been manipulated to eliminate all but one of the unknown fluctuating quantities, say $\tilde{q}^1$. The resulting equation will take the form

$$\sum_{n=0}^{N} a_n(s)\frac{d^n \tilde{q}^1}{ds^n} = 0 \tag{9.43}$$

In general, the coefficients $a_n(s)$, $n = 0 \to N$, will be complex functions of the mean flow and of the frequency. It follows that there are N independent solutions which, for all the independent fluctuating quantities, may be expressed in the form

$$\{\tilde{q}^n\} = [B(s)]\{A\} \tag{9.44}$$

where $[B(s)]$ is a matrix of complex solutions and $\{A\}$ is a vector of arbitrary complex constants to be determined from the boundary conditions. Consequently, the inlet and discharge fluctuations denoted by subscripts 1 and 2, respectively, are given by

$$\{\tilde{q}_1^n\} = [B(s_1)]\{A\}; \quad \{\tilde{q}_2^n\} = [B(s_2)]\{A\} \tag{9.45}$$

and therefore the transfer function

$$[T] = [B(s_2)][B(s_1)]^{-1} \tag{9.46}$$

Now for a uniform system, the coefficients a_n and the matrix $[B]$ are independent of s. Hence the equation 9.43 has a solution of the form

$$[B(s)] = [C][E] \tag{9.47}$$

where $[C]$ is a known matrix of constants, and $[E]$ is a diagonal matrix in which

$$E_{nn} = e^{j\gamma_n s} \tag{9.48}$$

where γ_n, $n = 1$ to N, are the solutions of the dispersion relation

$$\sum_{n=0}^{N} a_n \gamma^n = 0 \tag{9.49}$$

Note that γ_n are the wavenumbers for the N types of wave of frequency, ω, which can propagate through the uniform system. In general, each of these waves has a distinct wave speed, c_n, given by $c_n = -\omega/\gamma_n$. It follows from equations 9.47, 9.48, and 9.46 that the transfer matrix for a uniform distributed system must take the form

$$[T] = [C][E^*][C]^{-1} \qquad (9.50)$$

where $[E^*]$ is a diagonal matrix with

$$E_{nn}^* = e^{j\gamma_n\ell} \qquad (9.51)$$

and $\ell = s_2 - s_1$.

An important diagnostic property arises from the form of the transfer matrix, 9.50, for a uniform distributed system. The determinant, D_T, of the transfer matrix $[T]$ is

$$D_T = \exp\{j(\gamma_1 + \gamma_2 + \cdots + \gamma_N)\ell\} \qquad (9.52)$$

Thus the value of the determinant is related to the sum of the wavenumbers of the N different waves which can propagate through the uniform distributed system. Furthermore, if all the wavenumbers, γ_n, are purely real, then

$$|D_T| = 1 \qquad (9.53)$$

The property that the modulus of the determinant of the transfer function is unity will be termed "quasi-reciprocity" and will be discussed further below. Note that this will only be the case in the absence of wave damping when γ_n and c_n are purely real.

Turning now to another property, a system is said to be "reciprocal" if, in the matrix $[Z]$ defined by

$$\left\{\begin{matrix} \tilde{p}_1^T \\ \tilde{p}_2^T \end{matrix}\right\} = [Z]\left\{\begin{matrix} \tilde{m}_1 \\ -\tilde{m}_2 \end{matrix}\right\} \qquad (9.54)$$

the transfer impedances Z_{12} and Z_{21} are identical (see Brown 1967 for the generalization of this property in systems of higher order). This is identical to the condition that the determinant, D_T, of the transfer matrix $[T]$ be unity:

$$D_T = 1 \qquad (9.55)$$

We shall see that a number of commonly used components have transfer functions which are reciprocal. In order to broaden the perspective we have introduced the property of "quasi-reciprocity" to signify those components in which the modulus of the determinant is unity or

$$|D_T| = 1 \qquad (9.56)$$

We have already noted that uniform distributed components with purely real wavenumbers are quasi-reciprocal. Note that a uniform distributed component will only be reciprocal when the wavenumbers tend to zero, as, for example, in incompressible flows in which the wave propagation speeds tend to infinity.

By utilizing the results of section 9.9 we can conclude that any series or parallel combination of reciprocal components will yield a reciprocal system. Also a series combination of quasi-reciprocal components will be quasi-reciprocal. However it is *not* necessarily true that a parallel combination of quasi-reciprocal components is quasi-reciprocal.

An even more restrictive property than reciprocity is the property of "symmetry." A "symmetric" component is one that has identical dynamical properties when turned around so that the discharge becomes the inlet, and the directional convention of the flow variables is reversed (Brown 1967). Then, in contrast to the regular transfer matrix, $[T]$, the effective transfer matrix under these reversed circumstances is $[TR]$ where

$$\left\{ \begin{array}{c} \tilde{p}_1^T \\ -\tilde{m}_1 \end{array} \right\} = [TR] \left\{ \begin{array}{c} \tilde{p}_2^T \\ -\tilde{m}_2 \end{array} \right\} \tag{9.57}$$

and, comparing this with the definition 9.31, we observe that

$$TR_{11} = T_{22}/D_T; \quad TR_{12} = T_{12}/D_T$$
$$TR_{21} = T_{21}/D_T; \quad TR_{22} = T_{11}/D_T \tag{9.58}$$

Therefore symmetry, $[T] = [TR]$, requires

$$T_{11} = T_{22} \quad \text{and} \quad D_T = 1 \tag{9.59}$$

Consequently, in addition to the condition, $D_T = 1$, required for reciprocity, symmetry requires $T_{11} = T_{22}$.

As with the properties of reciprocity and quasi-reciprocity, it is useful to consider the property of a system comprised of symmetric components. Note that according to the combination rules of section 9.9, a parallel combination of symmetric components is symmetric, whereas a series combination may not retain this property. In this regard symmetry is in contrast to quasi-reciprocity in which the reverse is true.

In the case of uniform distributed systems, Brown (1967) shows that symmetry requires

$$F_{11} = F_{22} = 0 \tag{9.60}$$

so that the solution of the equation 9.40 for λ is $\lambda = \pm\lambda^*$ where $\lambda^* = (F_{21}F_{12})^{\frac{1}{2}}$ is known as the "propagation operator" and the transfer function 9.39 becomes

$$T_{11} = T_{22} = \cosh\lambda^*\ell$$
$$T_{12} = Z_C \sinh\lambda^*\ell$$
$$T_{21} = Z_C^{-1} \sinh\lambda^*\ell \tag{9.61}$$

where $Z_C = (F_{12}/F_{21})^{\frac{1}{2}} = (T_{12}/T_{21})^{\frac{1}{2}}$ is known as the "characteristic impedance."

In addition to the above properties of transfer functions, there are also properties associated with the net flux of fluctuation energy into the component or system. These will be elucidated after we have examined some typical transfer functions for components of hydraulic systems.

9.11 Some Simple Transfer Matrices

The flow of an incompressible fluid in a straight, rigid pipe will be governed by the following versions of equations 9.6 and 9.7:

$$\frac{\partial u}{\partial s} = 0 \tag{9.62}$$

$$\frac{\partial p^T}{\partial s} = -\frac{\rho f u|u|}{4a} - \rho\frac{\partial u}{\partial t} \tag{9.63}$$

If the velocity fluctuations are small compared with the mean velocity denoted by U (positive in direction from inlet to discharge), and the term $u|u|$ is linearized, then the above equations lead to the transfer function

$$[T] = \begin{bmatrix} 1 & -(R+j\omega L) \\ 0 & 1 \end{bmatrix} \tag{9.64}$$

where $(R+j\omega L)$ is the "impedance" made up of a "resistance," R, and an "inertance," L, given by

$$R = \frac{fU\ell}{2aA}, \quad L = \frac{\ell}{A} \tag{9.65}$$

where A, a, and ℓ are the cross-sectional area, radius, and length of the pipe. A number of different pipes in series would then have

$$R = Q\sum_i \frac{f_i\ell_i}{2a_i A_i^2}; \quad L = \sum_i \frac{\ell_i}{A_i} \tag{9.66}$$

where Q is the mean flow rate. For a duct of non-uniform area

$$R = Q \int_0^\ell \frac{f(s)ds}{2a(s)(A(s))^2}; \quad L = \int_0^\ell \frac{ds}{A(s)} \tag{9.67}$$

Note that all such ducts represent *reciprocal* and *symmetric* components.

A second, common hydraulic element is a simple "compliance," exemplified by an accumulator or a surge tank. It consists of a device installed in a pipeline and storing a volume of fluid, V_L, which varies with the local pressure, p, in the pipe. The compliance, C, is defined by

$$C = \rho \frac{dV_L}{dp} \tag{9.68}$$

In the case of a gas accumulator with a mean volume of gas, $\bar{V}_G$, which behaves according to the polytropic index, k, it follows that

$$C = \rho \bar{V}_G / k\bar{p} \tag{9.69}$$

where $\bar{p}$ is the mean pressure level. In the case of a surge tank in which the free surface area is A_S, it follows that

$$C = A_S / g \tag{9.70}$$

The relations across such compliances are

$$\tilde{m}_2 = \tilde{m}_1 - j\omega C \tilde{p}^T; \quad \tilde{p}_1^T = \tilde{p}_2^T = \tilde{p}^T \tag{9.71}$$

Therefore, using the definition 9.35, the transfer function $[T]$ becomes

$$[T] = \begin{bmatrix} 1 & 0 \\ -j\omega C & 1 \end{bmatrix} \tag{9.72}$$

Again, this component is *reciprocal* and *symmetric*, and is equivalent to a capacitor to ground in an electrical circuit.

Systems made up of lumped resistances, R, inertances, L, and compliances, C, will be termed LRC systems. Individually, all three of these components are both reciprocal and symmetric. It follows that any system comprised of these components will also be reciprocal (see section 9.10); hence all LRC systems are reciprocal. Note also that, even though individual components are symmetric, LRC systems are not symmetric since series combinations are not, in general, symmetric (see section 9.10).

An even more restricted class of systems are those consisting only of inertances, L, and compliances, C. These systems are termed "dissipationless" and have some special properties (see, for example, Pipes 1963) though these are rarely applicable in hydraulic systems.

As a more complicated example, consider the frictionless ($f = 0$) compressible flow in a straight uniform pipe of mean cross-sectional area, A_0. This can readily be shown to have the transfer function

$$T_{11}^* = (\cos\theta + jM\sin\theta)\,e^{j\theta M}$$

$$T_{12}^* = -j\bar{U}\sin\theta e^{j\theta M}\big/A_0 M$$

$$T_{21}^* = -jA_0 M(1-M^2)\sin\theta e^{j\theta M}\big/\bar{U}$$

$$T_{22}^* = (\cos\theta - jM\sin\theta)\,e^{j\theta M} \tag{9.73}$$

where $\bar{U}$ is the mean fluid velocity, $M = \bar{U}/c$ is the Mach number, and θ is a reduced frequency given by

$$\theta = \omega\ell/c(1 - M^2) \tag{9.74}$$

Note that all the usual acoustic responses can be derived quite simply from this transfer function. For example, if the pipe opens into reservoirs at both ends, so that appropriate inlet and discharge conditions are $\tilde{p}_1 = \tilde{p}_2 = 0$, then the transfer function, equation 9.35, can only be satisfied with $\tilde{m}_1 \neq 0$ if $T_{12}^* = 0$. According to equations 9.73, this can only occur if $\sin\theta = 0$, $\theta = n\pi$ or

$$\omega = n\pi c(1 - M^2)\big/\ell \tag{9.75}$$

which are the natural organ-pipe modes for such a pipe. Note also that the determinant of the transfer matrix is

$$D_T = D_{T^*} = e^{2j\theta M} \tag{9.76}$$

Since no damping has been included, this component is an undamped distributed system, and is therefore quasi-reciprocal. At low frequencies and Mach numbers, the transfer function 9.73 reduces to

$$T_{11}^* \to 1; \quad T_{12}^* \to -\frac{j\omega\ell}{A_0}$$

$$T_{21}^* \to -j\left(\frac{A_0\ell}{c^2}\right)\omega; \quad T_{22}^* \to 1 \tag{9.77}$$

and so consists of an inertance, ℓ/A_0, and a compliance, $A_0\ell/c^2$.

When friction is included (as is necessary in most water-hammer analyses) the transfer function becomes

$$T_{11}^* = \left(k_1 e^{k_1} - k_2 e^{k_2}\right) \Big/ (k_1 - k_2)$$

$$T_{12}^* = -\bar{U}(j\theta + f^*)\left(e^{k_1} - e^{k_2}\right) \Big/ A_0 M (k_1 - k_2)$$

$$T_{21}^* = -j\theta A_0 M (1 - M^2)\left(e^{k_1} - e^{k_2}\right) \Big/ \bar{U}(k_1 - k_2)$$

$$T_{22}^* = \left(k_1 e^{k_2} - k_2 e^{k_1}\right) \Big/ (k_1 - k_2) \tag{9.78}$$

in which $f^* = f\ell M/2a(1 - M^2)$ and k_1, k_2 are the solutions of

$$k^2 - kM(2j\theta + f^*) - j\theta(1 - M^2)(j\theta + f^*) = 0 \tag{9.79}$$

The determinant of this transfer matrix $[T^*]$ is

$$D_{T^*} = e^{k_1 + k_2} \tag{9.80}$$

Note that this component is only quasi-reciprocal in the undamped limit, $f \to 0$.

9.12 Fluctuation Energy Flux

It is clearly important to be able to establish the net energy flux into or out of a hydraulic system component (see Brennen and Braisted 1980). If the fluid is incompressible, and the order two system is characterized by the mass flow rate, m, and the total pressure, p^T, then the instantaneous energy flux through any system node is given by mp^T/ρ where the density is assumed constant. Substituting the expansions 9.28, 9.29 for p^T and m, it is readily seen that the mean flux of energy due to the fluctuations, E, is given by

$$E = \frac{1}{4\rho}\left\{\tilde{m}\bar{\tilde{p}}^T + \bar{\tilde{m}}\tilde{p}^T\right\} \tag{9.81}$$

where the overbar denotes a complex conjugate. Superimposed on E are fluctuations in the energy flux whose time-average value is zero, but we shall not be concerned with those fluctuations. The mean fluctuation energy flux, E, is of more consequence in terms, for example, of evaluating stability. It follows that the net flux of fluctuation energy *into* a component from the fluid is given by

$$E_1 - E_2 = \Delta E = \frac{1}{4\rho}\left[\tilde{m}_1\bar{\tilde{p}}_1^T + \bar{\tilde{m}}_1\tilde{p}_1^T - \tilde{m}_2\bar{\tilde{p}}_2^T - \bar{\tilde{m}}_2\tilde{p}_2^T\right] \tag{9.82}$$

and when the transfer function form 9.31 is used to write this in terms of the inlet fluctuating quantities

$$\Delta E = \frac{1}{4\rho} \left[-\Gamma_1 \tilde{p}_1^T \bar{\tilde{p}}_1^T - \Gamma_2 \tilde{m}_1 \bar{\tilde{m}}_1 + (1 - \Gamma_3) \tilde{m}_1 \bar{\tilde{p}}_1^T + (1 - \bar{\Gamma}_3) \bar{\tilde{m}}_1 \tilde{p}_1^T \right] \qquad (9.83)$$

where

$$\Gamma_1 = T_{11} \bar{T}_{21} + T_{21} \bar{T}_{11}$$
$$\Gamma_2 = T_{22} \bar{T}_{12} + T_{12} \bar{T}_{22}$$
$$\Gamma_3 = \bar{T}_{11} T_{22} + \bar{T}_{21} T_{12} \qquad (9.84)$$

and

$$|\Gamma_3|^2 = |D_T|^2 + \Gamma_1 \Gamma_2 \qquad (9.85)$$

Using the above relations, we can draw the following conclusions:

1. A component or system which is "conservative" (in the sense that $\Delta E = 0$ under all circumstances, whatever the values of $\tilde{p}_1^T$ and $\tilde{m}_1$) requires that

$$\Gamma_1 = \Gamma_2 = 0, \quad \Gamma_3 = 1 \qquad (9.86)$$

and these in turn require not only that the system or component be "quasi-reciprocal" ($|D_T| = 1$) but also that

$$\frac{\bar{T}_{11}}{T_{11}} = -\frac{\bar{T}_{12}}{T_{12}} = -\frac{\bar{T}_{21}}{T_{21}} = \frac{\bar{T}_{22}}{T_{22}} = \frac{1}{D_T} \qquad (9.87)$$

Such conditions virtually never occur in real hydraulic systems, though any combination of lumped inertances and compliances does constitute a conservative system. This can be readily demonstrated as follows. An inertance or compliance has $D_T = 1$, purely real T_{11} and T_{22} so that $T_{11} = \bar{T}_{11}$ and $T_{22} = \bar{T}_{22}$, and purely imaginery T_{21} and T_{12} so that $T_{21} = -\bar{T}_{21}$ and $T_{12} = -\bar{T}_{12}$. Hence individual inertances or compliances satisfy equations 9.86 and 9.87. Furthermore, from the combination rules of section 9.9, it can readily be seen that all combination of components with purely real T_{11} and T_{22} and purely imaginery T_{21} and T_{12} will retain the same properties. Consequently, any combination of inertance and compliance satisfies equations 9.86 and 9.87 and is conservative.
2. A component or system will be considered "completely passive" if ΔE is positive for all possible values of $\tilde{m}_1$ and $\tilde{p}_1^T$. This implies that a net external supply of energy to the fluid is required to maintain any steady state oscillation. To find

the characteristics of the transfer function which imply "complete passivity" the expression 9.83 is rewritten in the form

$$\Delta E = \frac{|\tilde{p}_1^T|^2}{4\rho}\left[-\Gamma_1 - \Gamma_2 x\bar{x} + (1 - \Gamma_3)x + (1 - \bar{\Gamma}_3)\bar{x}\right] \qquad (9.88)$$

where $x = \tilde{m}_1/\tilde{p}_1^T$. It follows that the sign of ΔE is determined by the sign of the expression in the square brackets. Moreover, if $\Gamma_2 < 0$, it is readily seen that this expression has a minimum and is positive for all x if

$$\Gamma_1\Gamma_2 > |1 - \Gamma_3|^2 \qquad (9.89)$$

which, since $\Gamma_2 < 0$, implies $\Gamma_1 < 0$. It follows that necessary and sufficient conditions for a component or system to be completely passive are

$$\Gamma_1 < 0 \quad \text{and} \quad G < 0 \qquad (9.90)$$

where

$$G = |1 - \Gamma_3|^2 - \Gamma_1\Gamma_2 = |D_T|^2 + 1 - 2Re\{\Gamma_3\} \qquad (9.91)$$

The conditions 9.90 also imply $\Gamma_2 < 0$. Conversely a "completely active" component or system which always has $\Delta E < 0$ occurs if and only if $\Gamma_1 > 0$ and $G < 0$ which imply $\Gamma_2 > 0$. These properties are not, of course, the only possibilities. A component or system which is not completely passive or active could be "potentially active." That is to say, ΔE could be negative for the right combination of $\tilde{m}_1$ and $\tilde{p}_1^T$, which would, in turn, depend on the rest of the system to which the particular component or system is attached. Since Γ_1 is almost always negative, it transpires that most components are either completely passive or potentially active, depending on the sign of the quantity, G, which will therefore be termed the "dynamic activity." These circumstances can be presented graphically as shown in figure 9.3.

   In practice, of course, both the transfer function, and properties like the dynamic activity, G, will be functions not only of frequency but also of the mean flow conditions. Hence the potential for system instability should be evaluated by tracking the graph of G against frequency, and establishing the mean flow conditions for which the quantity G becomes negative within the range of frequencies for which transfer function information is available.

   While the above analysis represents the most general approach to the stability of systems or components, the results are not readily interpreted in terms of commonly employed measures of the system or component characteristics. It is therefore instructive to consider two special subsets of the general case, not only because of the

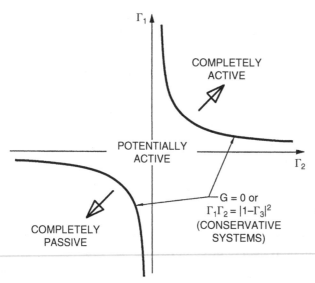

Figure 9.3. Schematic of the conditions for completely active, completely passive and potentially active components or systems.

simplicity of the results, but also because of the ubiquity of these special cases. Consider first a system or component that discharges into a large, constant head reservoir, so that $\tilde{p}_2^T = 0$. It follows from the expression 9.82 that

$$\Delta E = \frac{|\tilde{m}_1|^2}{2\rho} Re\{\tilde{p}_1^T / \tilde{m}_1\} \tag{9.92}$$

Note that ΔE is always purely real and that the sign only depends on the real part of the "input impedance"

$$\tilde{p}_1^T / \tilde{m}_1 = -T_{12}/T_{11} \tag{9.93}$$

Consequently a component or system with a constant head discharge will be dynamically stable if the "input resistance" is positive or

$$Re\left\{-T_{12}/T_{11}\right\} > 0 \tag{9.94}$$

This relation between the net fluctuation energy flux, the input resistance, and the system stability, is valuable because of the simplicity of its physical interpretation. In practice, the graph of input resistance against frequency can be monitored for changes with mean flow conditions. Instabilities will arise at frequencies for which the input resistance becomes negative.

The second special case is that in which the component or system begins with a constant head reservoir rather than discharging into one. Then

$$\Delta E = \frac{|\tilde{m}_2|^2}{2\rho} Re\left\{-\tilde{p}_2^T / \tilde{m}_2\right\} \tag{9.95}$$

and the stability depends on the sign of the real part of the "discharge impedance"

$$-\tilde{p}_2^T / \tilde{m}_2 = -T_{12}/T_{22} \tag{9.96}$$

Thus a constant head inlet component or system will be stable when the "discharge resistance" is positive or

$$Re\left\{-T_{12}/T_{22}\right\} > 0 \tag{9.97}$$

In practice, since T_{11} and T_{22} are close to unity for many components and systems, both the condition 9.94 and the condition 9.97 reduce to the approximate condition that the system resistance, $Re\{-T_{12}\}$, be positive for system stability. While not always the case, this approximate condition is frequently more convenient and more readily evaluated than the more precise conditions detailed above and given in equations 9.94 and 9.97. Note specifically, that the system resistance can be obtained from steady state operating characteristics; for example, in the case of a pump or turbine, it is directly related to the slope of the head-flow characteristic and instabilities in these devices which result from operation in a regime where the slope of the characteristic is positive and $Re\{-T_{12}\}$ is negative are well known (Greitzer 1981) and have been described earlier (section 8.6).

It is, however, important to recognize that the approximate stability criterion $Re\{-T_{12}\} > 0$, while it may provide a useful guideline in many circumstances, is by no means accurate in all cases. One notable and important case in which this criterion is inaccurate is the auto-oscillation phenomenon described in section 8.7. This is *not* the result of a positive slope in the head-flow characteristic, but rather occurs where this slope is negative and is caused by cavitation-induced changes in the other elements of the transfer function. This circumstance will be discussed further in section 9.14.

9.13 Non-Cavitating Pumps

Consider now the questions associated with transfer functions for pumps or other turbomachines. In the simple fluid flows of section 9.11 we were able to utilize the known equations governing the flow in order to construct the transfer functions for those simple components. In the case of more complex fluids or geometries, one cannot necessarily construct appropriate one-dimensional flow equations, and therefore must resort to results derived from more global application of conservation laws or to experimental measurements of transfer matrices. Consider first the transfer matrix, $[TP]$, for incompressible flow through a pump (all pump transfer functions will be of the $[T]$ form defined in equation 9.35) which will clearly be a function not only of the frequency, ω, but also of the mean operating point as represented by the flow coefficient, ϕ, and the cavitation number, σ. At very low frequencies one can argue that the pump will simply track up and down the performance characteristic, so that, for small amplitude perturbations and in the absence of cavitation, the transfer function becomes

$$[TP] = \begin{bmatrix} 1 & \frac{d(\Delta p^T)}{dm} \\ 0 & 1 \end{bmatrix} \qquad (9.98)$$

where $d(\Delta p^T)/dm$ is the slope of the steady state operating characteristic of total pressure rise versus mass flow rate. Thus we define the pump resistance, $R_P = -d(\Delta p^T)/dm$, where R_P is usually positive under design flow conditions, but may be negative at low flow rates as discussed earlier (section 8.6). At finite frequencies, the elements TP_{21} and TP_{22} will continue to be zero and unity respectively, since the instantaneous flow rate into and out of the pump must be identical when the fluid and structure are incompressible and no cavitation occurs. Furthermore, TP_{11} must continue to be unity since, in an incompressible flow, the total pressure differences must be independent of the level of the pressure. It follows that the transfer function at higher frequencies will become

$$[TP] = \begin{bmatrix} 1 & -I_P \\ 0 & 1 \end{bmatrix} \qquad (9.99)$$

where the pump impedence, I_P, will, in general, consist of a resistive part, R_P, and a reactive part, $j\omega L_P$. The resistance, R_P, and inertance, L_P, could be functions of both the frequency, ω, and the mean flow conditions. Such simple impedance models for pumps have been employed, together with transfer functions for the suction and discharge lines (equation 9.73), to model the dynamics of pumping systems. For example, Dussourd (1968) used frequency domain methods to analyse pulsation problems in boiler feed pump systems. More recently, Sano (1983) used transfer functions to obtain natural frequencies for pumping systems that agree well with those observed experimentally.

The first fundamental investigation of the dynamic response of pumps seems to have been carried out by Ohashi (1968) who analyzed the oscillating flow through a cascade, and carried out some preliminary experimental investigations on a centrifugal pump. These studies enabled him to evaluate the frequency at which the response of the pump would cease to be quasistatic (see below). Fanelli (1972) appears to have been the first to explore the nature of the pump transfer function, while the first systematic measurements of the impedance of a noncavitating centrifugal pump are those of Anderson, Blade, and Stevans (1971). Typical resistive and reactive component measurements from the work of Anderson, Blade, and Stevans are reproduced in figure 9.4. Note that, though the resistance approaches the quasistatic value at low frequencies, it also departs significantly from this value at higher frequencies. Moreover, the reactive part is only roughly linear with frequency. The resistance and inertance are presented again in figure 9.5, where they are compared with the results of a dynamic

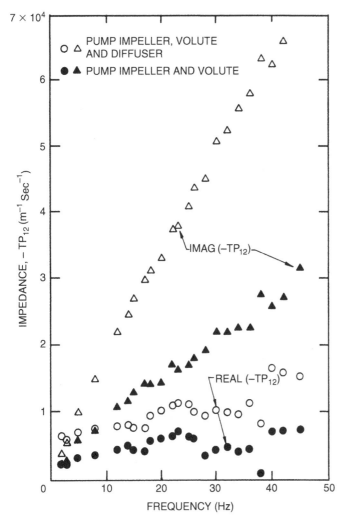

Figure 9.4. Impedance measurements made by Anderson, Blade, and Stevans (1971) on a centrifugal pump (impeller diameter of $18.9\ cm$) operating at a flow coefficient of 0.442 and a speed of 3000 rpm. The real or resistive part of $(-T_{12})$ and the imaginary or reactive part of (T_{12}) are plotted against the frequency of the perturbation.

model proposed by Anderson, Blade and Stevans. In this model, each pump impeller passage is represented by a resistance and an inertance, and the volute by a series of resistances and inertances. Since each impeller passage discharges into the volute at different locations relative to the volute discharge, each impeller passage flow experiences a different impedance on its way to the discharge. This results in an overall pump resistance and inertance that are frequency dependent as shown in figure 9.5. Note that the comparison with the experimental observations (which are also included in figure 9.5) is fair, but not completely satisfactory. Moreover, it should be noted, that the comparison shown is for a flow coefficient of 0.442 (above the design flow

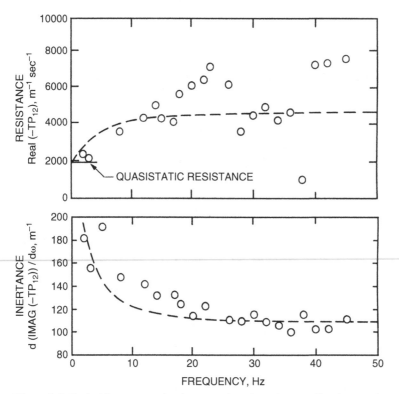

Figure 9.5. Typical inertance and resistance values from the centrifugal pump data of figure 9.4. Data do not include the diffuser contribution. The lines correspond to analytical values obtained as described in the text.

coefficient), and that, at higher flow coefficients, the model and experimental results exhibited poorer agreement.

Subsequent measurements of the impedance of non-cavitating axial and mixed flow pumps by Ng and Brennen (1978) exhibit a similar increase in the resistance with frequency (see next section). In both sets of dynamic data, it does appear that significant departure from the quasistatic values can be expected when the reduced frequency, (frequency/rotation frequency) exceeds about 0.02 (see figures 9.5 and 9.6). This is roughly consistent with the criterion suggested by Ohashi (1968) who concluded that non-quasistatic effects would occur above a reduced frequency of $0.05 Z_R \phi / \cos \beta$. For the inducers of Ng and Brennen, Ohashi's criterion yields values for the critical reduced frequency of about 0.015.

9.14 Cavitating Inducers

In the presence of cavitation, the transfer function for a pump or inducer will be considerably more complicated than that of equation 9.99. Even at low frequencies, the values of $T P_{11}$ will become different from unity, because the head rise will change with the inlet total pressure, as manifest by the nonzero value of $d(\Delta p^T)/dp_1^T$ at a

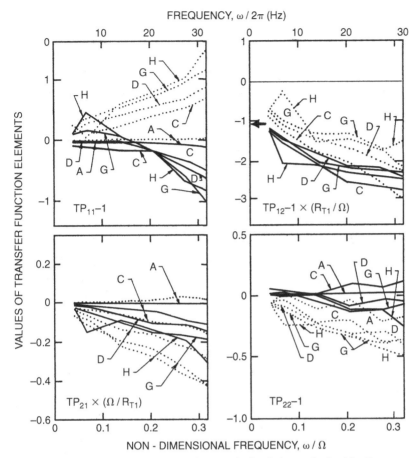

Figure 9.6. Typical transfer functions for a cavitating inducer obtained by Brennen *et al.* (1982) for a 10.2 *cm* diameter inducer (Impeller VI) operating at 6000 *rpm* and a flow coefficient of $\phi_1 = 0.07$. Data is shown for four different cavitation numbers, σ = (A) 0.37, (C) 0.10, (D) 0.069, (G) 0.052, and (H) 0.044. Real and imaginary parts are denoted by the solid and dashed lines respectively. The quasistatic pump resistance is indicated by the arrow (adapted from Brennen *et al.* 1982).

given mass flow rate, m_1. Furthermore, the volume of cavitation, $V_C(p_1^T, m_1)$, will vary with both the inlet total pressure, p_1^T (or $NPSH$ or cavitation number), and with the mass flow rate, m_1 (or with angle of incidence), so that

$$[TP] = \begin{bmatrix} 1 + \frac{d(\Delta p^T)}{dp_1^T}|_{m_1} & \frac{d\Delta p^T}{dm_1}|_{p_1^T} \\[2em] j\omega\rho_L \frac{dV_C}{dp_1^T}|_{m_1} & 1 + j\omega\rho_L \frac{dV_C}{dm_1}|_{p_1^T} \end{bmatrix} \qquad (9.100)$$

Brennen and Acosta (1973, 1975, 1976) identified this quasistatic or low frequency form for the transfer function of a cavitating pump, and calculated values of the cavitation compliance, $-\rho_L(dV_C/dp_1^T)_{m_1}$ and the cavitation mass flow gain factor, $-\rho_L(dV_C/dm_1)_{p_1^T}$, using the cavitating cascade solution discussed in section 7.10.

Both the upper limit of frequency at which this quasistatic approach is valid and the form of the transfer function above this limit cannot readily be determined except by experiment. Though it was clear that experimental measurements of the dynamic transfer functions were required, these early investigations of Brennen and Acosta did highlight the importance of both the compliance and the mass flow gain factor in determining the stability of systems with cavitating pumps.

Ng and Brennen (1978) and Brennen $et\ al.$ (1982) conducted the first experiments to measure the complete transfer function for cavitating inducers. Typical transfer functions are those for the 10.2 cm diameter Impeller VI (see section 2.8), whose noncavitating steady state performance was presented in figure 7.15. Transfer matrices for that inducer are presented in figure 9.6 as a function of frequency (up to 32 Hz), for a speed of 6000 rpm, a flow coefficient $\phi_1 = 0.07$ and for five different cavitation numbers ranging from data set A that was taken under noncavitating conditions, to data set C that showed a little cavitation, to data set H that was close to breakdown. The real and imaginary parts are represented by the solid and dashed lines, respectively. Note, first, that, in the absence of cavitation (Case A), the transfer function is fairly close to the anticipated form of equation 9.99 in which $TP_{11} = TP_{22} = 1$, $TP_{21} = 0$. Also, the impedance (TP_{12}) is comprised of an expected inertance (the imaginary part of TP_{12} is linear in frequency) and a resistance (real part of $-TP_{12}$) which is consistent with the quasistatic resistance from the slope of the head rise characteristic (shown by the arrow in figure 9.6 at $TP_{12}R_{T1}/\Omega = 1.07$). The resistance appears to increase with increasing frequency, a trend which is consistent with the centrifugal pump measurements of Anderson, Blade and Stevans (1971) which were presented in figure 9.5.

It is also clear from figure 9.6 that, as the cavitation develops, the transfer function departs significantly from the form of equation 9.99. One observes that TP_{11} and TP_{22} depart from unity, and develop nonzero imaginary parts that are fairly linear with frequency. Also TP_{21} becomes nonzero, and, in particular, exhibits a compliance which clearly increases with decreasing cavitation number. All of these changes mean that the determinant, D_{TP}, departs from unity as the cavitation becomes more extensive. This is illustrated in figure 9.7, which shows the determinant corresponding to the data of figure 9.6. Note that $D_{TP} \approx 1$ for the non-cavitating case A, but that it progressively deviates from unity as the cavitation increases. We can conclude that the presence of cavitation can cause a pump to assume potentially active dynamic characteristics when it would otherwise be dynamically passive.

Polynomials of the form

$$TP_{ij} = \sum_{n=0}^{n^*} A_{nij}(j\omega)^n \qquad (9.101)$$

were fitted to the experimental transfer function data using values of n^* of 3 or 5. To illustrate the result of such curve fitting we include figure 9.8, which depicts the result

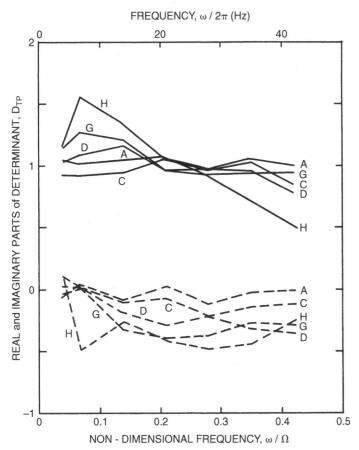

Figure 9.7. Determinant, D_{TP}, of the experimental transfer functions of figure 9.6. The real and imaginary parts are shown by the solid and dashed lines respectively and, as in figure 9.6, the letter code A→H refers to steady state operating points with increasing cavitation (adapted from Brennen *et al.* 1982).

of curve fitting figure 9.6. We now proceed to examine several of the coefficients A_{nij} that are of particular interest (note that $A_{011} = A_{022} = 1$, $A_{021} = 0$ for reasons described earlier). We begin with the inertance, $-A_{112}$, which is presented nondimensionally in figure 9.9. Though there is significant scatter at the lower cavitation numbers, the two different sizes of inducer pump appear to yield similar inertances. Moreover, the data suggest some decrease in the inertance with decreasing σ. On the other hand, the corresponding data for the compliance, $-A_{121}$, which is presented in figure 9.10 seems roughly inversely proportional to the cavitation number. And the same is true for both the mass flow gain factor, $-A_{122}$, and the coefficient that defines the slope of the imaginary part of TP_{11}, A_{111}; these are presented in figures 9.11 and 9.12, respectively. All of these data appear to conform to the physical scaling implicit in the nondimensionalization of each of the dynamic characteristics.

It is also valuable to consider the results of figures 9.9 to 9.12 in the context of an analytical model for the dynamics of cavitating pumps (Brennen 1978). We present

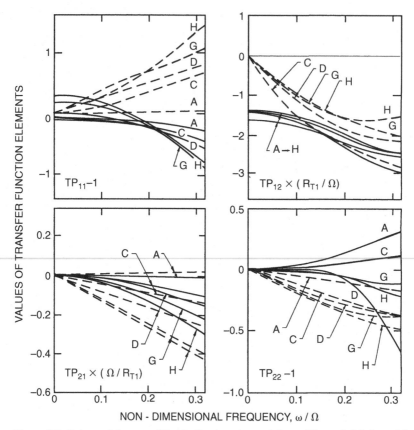

Figure 9.8. Polynomial curves fitted to the experimental data of figure 9.6 (adapted from Brennen *et al.* 1982).

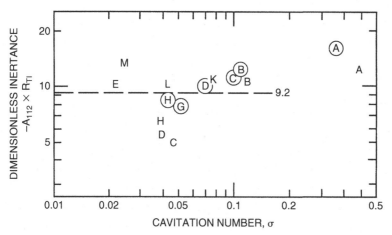

Figure 9.9. The inertance, $-A_{112}$, non-dimensionalized as $-A_{112}R_{T1}$, as a function of cavitation number for two axial inducer pumps (Impellers IV and VI) with the same geometry but different diameters. Data for the 10.2 *cm* diameter Impeller VI is circled and was obtained from the data of figure 9.6. The uncircled points are for the 7.58 *cm* diameter Impeller IV. Adapted from Brennen *et al.* (1982).

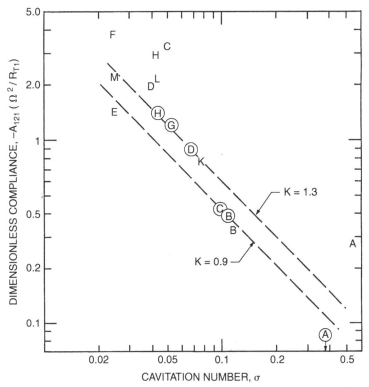

Figure 9.10. The compliance, $-A_{121}$, nondimensionalized as $-A_{121}\Omega^2/R_{T1}$ for the same circumstances as described in figure 9.9.

here a brief physical description of that model, the essence of which is depicted schematically in figure 9.13, which shows a developed, cylindrical surface within the inducer. The cavitation is modeled as a bubbly mixture which extends over a fraction, ϵ, of the length, c, of each blade passage before collapsing at a point where the pressure has risen to a value which causes collapse. The mean void fraction of the bubbly mixture is denoted by α_0. Thus far we have described a flow which is nominally steady. We must now consider perturbing both the pressure and the flow rate at inlet, since the relation between these perturbations, and those at discharge, determine the transfer function. Pressure perturbations at inlet will cause pressure waves to travel through the bubbly mixture and this part of the process is modeled using a mixture compressibility parameter, K, to determine that wave speed. On the other hand, fluctuations in the inlet flow rate produce fluctuations in the angle of incidence which cause fluctuations in the rate of production of cavitation at inlet. These disturbances would then propagate down the blade passage as kinematic or concentration waves which travel at the mean mixture velocity. This process is modeled by a factor of proportionality, M, which relates the fluctuation in the angle of incidence to the fluctuations in the void fraction. Neither of the parameters, K or M, can be readily estimated analytically; they are, however, the two key features in the bubbly flow

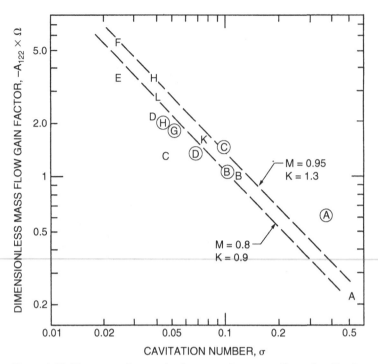

Figure 9.11. The mass flow gain factor, $-A_{122}$, nondimensionalized as $-A_{122}\Omega$ for the same circumstances as described in figure 9.9.

model. Moreover they respectively determine the cavitation compliance and the mass flow gain factor, two of the most important factors in the transfer function insofar as the prediction of instability is concerned.

The theory yields the following expressions for A_{111}, A_{112}, A_{121}, and A_{122} at small dimensionless frequencies (Brennen 1978, 1982):

$$A_{111}\Omega \simeq \frac{K\zeta\epsilon}{4}\left\{\cot\beta_{b1} + \phi_1/\sin^2\beta_{b1}\right\}$$

$$A_{112}R_{T1} \simeq -\zeta/4\pi\sin^2\beta_{b1}$$

$$A_{121}\Omega^2/R_{T1} \simeq -\pi K\zeta\epsilon/4$$

$$A_{122}\Omega \simeq -\frac{\zeta\epsilon}{4}\left\{M/\phi_1 - K\phi_1/\sin^2\beta_{b1}\right\} \tag{9.102}$$

where $\zeta = \ell Z_R/R_{T1}$ where ℓ is the axial length of the inducer, and Z_R is the number of blades. Evaluation of the transfer function elements can be effected by noting that the experimental observations suggest $\epsilon \approx 0.02/\sigma$. Consequently, the A_{nij} characteristics from equations 9.102 can be plotted against cavitation number. Typical results are shown in figures 9.9 through 9.12 for various choices of the two undetermined parameters K and M. The inertance, A_{112}, which is shown in figure 9.9, is independent of K and M. The calculated value of the inertance for these impellers is about 9.2; the

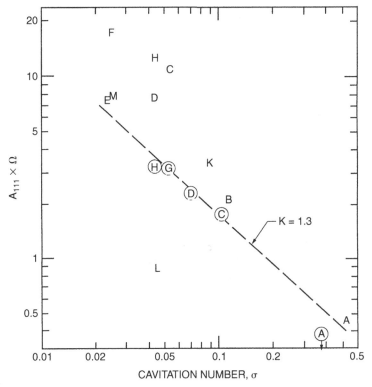

Figure 9.12. The characteristic, A_{111}, nondimensionalized as $A_{111}\Omega$ for the same circumstances as described in figure 9.9.

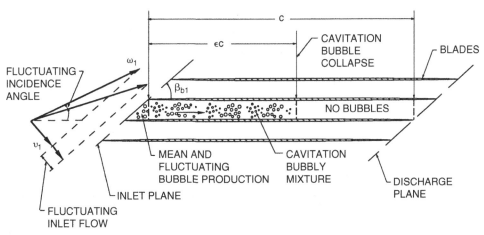

Figure 9.13. Schematic of the bubbly flow model for the dynamics of cavitating pumps (adapted from Brennen 1978).

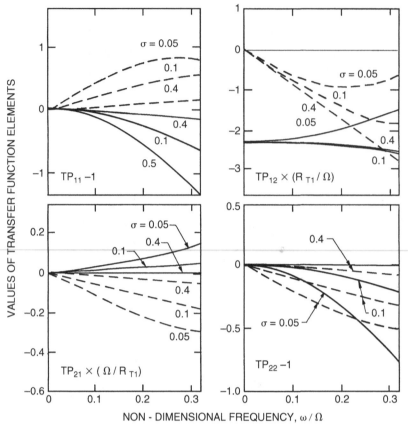

Figure 9.14. Transfer functions for Impellers VI and IV at $\phi_1 = 0.07$ calculated from the bubbly flow model using $K = 1.3$ and $M = 0.8$ (adapted from Brennen *et al.* 1982).

actual value may be somewhat larger because of three-dimensional geometric effects that were not included in the calculation (Brennen *et al.* 1982). The parameter M only occurs in A_{122}, and it appears from figure 9.11 as though values of this parameter in the range $0.8 \rightarrow 0.95$ provide the best agreement with the data. Also, a value of $K \approx 1.3$ seems to generate a good match with the data of figures 9.10, 9.11, and 9.12.

Finally, since $K = 1.3$ and $M = 0.8$ seem appropriate values for these impellers, we reproduce in figure 9.14 the complete theoretical transfer functions for various cavitation numbers. These should be directly compared with the transfer functions of figure 9.8. Note that the general features of the transfer functions, and their variation with cavitation number, are reproduced by the model. The most notable discrepency is in the real part of TP_{21}; this parameter is, however, usually rather unimportant in determining the stability of a hydraulic system. Most important from the point of view of stability predictions, the cavitation compliance and mass flow gain factor components of the transfer function are satisfactorily modeled.

9.15 System with Rigid Body Vibration

All of the preceding analysis has assumed that the structure of the hydraulic system is at rest in some inertial coordinate system. However, there are a number of important problems in which the oscillation of the hydraulic system itself may play a central role. For instance, one might seek to evaluate the unsteady pressures and flow rates in a hydraulic system aboard a vehicle undergoing translational or rotational oscillations. Examples might be oil or water pumping systems aboard a ship, or the fuel and hydraulic systems on an aircraft. In other circumstances, the motion of the vehicle may couple with the propulsion system dynamics to produce instabilities, as in the simplest of the Pogo instabilities of liquid propelled rocket engines (see section 8.13).

In this section we give a brief outline of how rigid body oscillations of the hydraulic system can be included in the frequency domain methodology. For convenience we shall refer to the structure of the hydraulic system as the "vehicle." There are, of course, more complex problems in which the deformation of the vehicle is important. Such problems require further refinement of the methods presented here.

In order to include the rigid body oscillation of the vehicle in the analysis, it is first necessary to define a coordinate system, $\underline{x}$, which is fixed in the vehicle, and a separate inertial or nonaccelerating coordinate system, $\underline{x}_A$. The mean location of the origin of the $\underline{x}$ system is chosen to coincide with the origin of the $\underline{x}_A$ system. The oscillations of the vehicle are then described by stating that the translational and rotational displacements of the $\underline{x}$ coordinate system in the $\underline{x}_A$ system are respectively given by

$$Re\left\{\underline{\tilde{d}}e^{j\omega t}\right\}; \quad Re\left\{\underline{\tilde{\theta}}e^{j\omega t}\right\} \tag{9.103}$$

It follows that the oscillatory displacement of any vector point, $\underline{x}$, in the vehicle is given by

$$Re\left\{(\underline{\tilde{d}}+\underline{\tilde{\theta}}\times\underline{x})e^{j\omega t}\right\} \tag{9.104}$$

and the oscillatory velocity of that point will be

$$Re\left\{j\omega(\underline{\tilde{d}}+\underline{\tilde{\theta}}\times\underline{x})e^{j\omega t}\right\} \tag{9.105}$$

Then, if the steady and oscillatory velocities of the flow in the hydraulic system, and *relative to* that system, are given as in the previous sections by $\underline{\bar{u}}$ and $\underline{\tilde{u}}$ respectively, it follows that the oscillatory velocity of the flow in the nonaccelerating frame, $\underline{\tilde{u}}_A$, is given by

$$\underline{\tilde{u}}_A = \underline{\tilde{u}} + j\omega(\underline{\tilde{d}}+\underline{\tilde{\theta}}\times\underline{x}) \tag{9.106}$$

Furthermore, the acceleration of the fluid in the nonaccelerating frame, $\tilde{\underline{a}}_A$, is given by

$$\tilde{\underline{a}}_A = j\omega\tilde{\underline{u}} - \omega^2\tilde{\underline{d}} - \omega^2\tilde{\underline{\theta}} \times \underline{x} + 2j\omega\tilde{\underline{\theta}} \times \tilde{\underline{u}} \qquad (9.107)$$

The last three terms on the right-hand side are vehicle-induced accelerations of the fluid in the hydraulic system. It follows that these accelerations will alter the difference in the total pressure between two nodes of the hydraulic system denoted by subscripts 1 and 2. By integration one finds that the total pressure difference, $(\tilde{p}_2^T - \tilde{p}_1^T)$, is related to that which would pertain in the absence of vehicle oscillation, $(\tilde{p}_2^T - \tilde{p}_1^T)_0$, by

$$(\tilde{p}_2^T - \tilde{p}_1^T) = (\tilde{p}_2^T - \tilde{p}_1^T)_0 + \rho\omega^2 \left\{ (\underline{x}_2 - \underline{x}_1) \cdot \tilde{\underline{d}} + (\underline{x}_2 \times \underline{x}_1) \cdot \tilde{\underline{\theta}} \right\} \qquad (9.108)$$

where $\underline{x}_2$ and $\underline{x}_1$ are the locations of the two nodes in the frame of reference of the vehicle.

The inclusion of these acceleration-induced total pressure changes is the first step in the synthesis of models of this class of problems. Their evaluation requires the input of the location vectors, $\underline{x}_i$, for each of the system nodes, and the values of the system displacement frequency, ω, and amplitudes, $\tilde{\underline{d}}$ and $\tilde{\underline{\theta}}$. In an analysis of the response of the hydraulic system, the vibration amplitudes, $\tilde{\underline{d}}$ and $\tilde{\underline{\theta}}$, would be included as inputs. In a stability analysis, they would be initially unknown. In the latter case, the system of equations would need to be supplemented by those of the appropriate feedback mechanism. An example would be a set of equations giving the unsteady thrust of an engine in terms of the fluctuating fuel supply rate and pressure and giving the accelerations of the vehicle resulting from that fluctuating thrust. Clearly a complete treatment of such problems would be beyond the scope of this book.

10

Radial and Rotordynamic Forces

10.1 Introduction

This chapter is devoted to a discussion of the various fluid-induced radial and rotor-dynamic forces which can occur in pumps and other turbomachines. It has become increasingly recognized that the reliability and acceptability of modern turbomachines depend heavily on the degree of vibration and noise which those machines produce (Makay and Szamody 1978), and that one of the most common sources of vibration is associated with the dynamics of the shaft and its related components, bearings, seals, and impellers (Duncan 1966–67, Doyle 1980, Ehrich and Childs 1984). It is clear that the modern pump designer (see, for example, Ek 1978, France 1986), or turbine designer (see, for example, Pollman *et al.* 1978), must pay particular attention to the rotordynamics of the shaft to ensure not only that the critical speeds occur at expected rotational rates, but also that the vibration levels are minimized. It is, however, important to note that not all shaft vibrations are caused by rotordynamic instability. For example, Rosenmann (1965) reports oscillating radial forces on cavitating inducers that are about 20% of the axial thrust, and are caused by flow oscillations, not rotordynamic oscillations. Also, Marscher (1988) investigated shaft motions induced by the unsteady flows at inlet to a centrifugal impeller operating below the design flow rate.

Texts such as Vance (1988) provide background on the methods of rotordynamic analysis. We focus here only on some of the inputs which are needed for that analysis, namely the forces caused by fluid motion in the bearing, seal, or impeller. One reason for this emphasis is that these inputs represent, at present, the area of greatest uncertainty insofar as the rotordynamic analysis is concerned.

We shall attempt to present data from many different sources using a common notation and a common nondimensionalizing procedure. This background is reviewed in the next section. Subsequently, we shall examine the known fluid-induced rotordynamic effects in hydrodynamic bearings, seals, and other devices. Then the forces acting on an impeller, both steady radial and rotordynamic forces, will be reviewed both for centrifugal pumps and for axial flow inducers.

10.2 Notation

The forces that the fluid imparts to the rotor in a plane perpendicular to the axis of rotation are depicted in figure 10.1, and are decomposed into components in the directions x and y, where this coordinate system is fixed in the framework of the pump. The instantaneous forces are denoted by $F_x^*(t)$, $F_y^*(t)$, and the time-averaged values of these forces in the stationary frame are denoted by F_{0x}^*, F_{0y}^*. By definition, these are the steady forces commonly referred to as the radial forces or radial thrust. Sometimes it is important to know the axial position of the line of action of these forces. Alternatively, one can regard the x, y axes as fixed at some convenient axial location. Then, in addition to the forces, $F_x^*(t)$ and $F_y^*(t)$, the fluid-induced bending moments, $M_x^*(t)$ and $M_y^*(t)$, would be required information. The time-averaged moments will be defined by M_{0x}^* and M_{0y}^*.

Even if the location of the center of rotation were stationary at the origin of the xy plane (figure 10.1) the forces $F_x^*(t)$, $F_y^*(t)$ and moments $M_x^*(t)$, $M_y^*(t)$ could still have significant unsteady components. For example, rotor-stator interaction could lead to significant forces on the impeller at the blade passing frequencies. Similarly, there could be blade passing frequency components in the torque, $T(t)$, and the axial thrust, as discussed earlier in section 8.2. For simplicity, however, they will not be included in the present mathematical formulation.

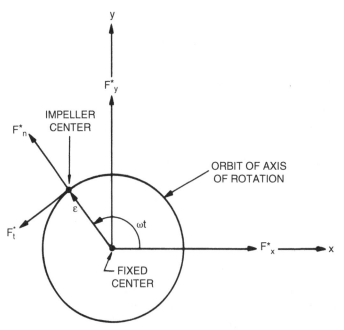

Figure 10.1. Schematic showing the relationship between the forces in the pump frame, F_x^*, F_y^*, the rotordynamic forces, F_n^*, F_t^*, the impeller center, the whirl orbit, and the volute geometry.

The other set of forces with which this chapter will be concerned are the fluid-induced rotordynamic forces that are caused by the displacement and motion of the axis of rotation. It will be assumed that this displacement is sufficiently small so that a linear perturbation model is accurate. Then

$$\left\{ \begin{array}{c} F_x^*(t) \\ F_y^*(t) \end{array} \right\} = \left\{ \begin{array}{c} F_{0x}^* \\ F_{0y}^* \end{array} \right\} + [A^*] \left\{ \begin{array}{c} x(t) \\ y(t) \end{array} \right\} \qquad (10.1)$$

where the displacement is given by $x(t)$ and $y(t)$, and $[A^*]$ is known as the "rotordynamic force matrix," which, in the linear model, would be independent of time, t. In virtually all cases that we shall be describing here, the displacements are sinusoidal. The "whirl" frequency of these motions will be denoted by ω (rad/s). Then, in general, the matrix $[A^*]$ will not only be a function of the turbomachine geometry and operating condition, but also of the whirl frequency, ω. In an analogous manner the rotordynamic moment matrix, $[B^*]$, is defined by

$$\left\{ \begin{array}{c} M_x^*(t) \\ M_y^*(t) \end{array} \right\} = \left\{ \begin{array}{c} M_{0x}^* \\ M_{0y}^* \end{array} \right\} + [B^*] \left\{ \begin{array}{c} x(t) \\ y(t) \end{array} \right\} \qquad (10.2)$$

The radial forces will be presented here in nondimensional form (denoted by the same symbols without the asterisk) by dividing the forces by $\rho \pi \Omega^2 R_{T2}^3 L$, where the selected length L may vary with the device. In seals and bearings, L is the axial length of the component. For centrifugal pumps, it is appropriate to use the width of the discharge so that $L = B_2$. With axial inducers, the axial extent of the blades is used for L. The displacements are nondimensionalized by R. In seals and bearings, the radius of the rotor is used; in centrifugal pump impellers, the discharge radius is used so that $R = R_{T2}$. It follows that the matrix $[A]$ is nondimensionalized by $\rho \pi \Omega^2 R^2 L$. Correspondingly, the radial moments and the moment matrix $[B]$ are nondimensionalized by $\rho \pi \Omega^2 R^4 L$ and $\rho \pi \Omega^2 R^3 L$ respectively. Thus

$$\left\{ \begin{array}{c} F_x(t) \\ F_y(t) \end{array} \right\} = \left\{ \begin{array}{c} F_{0x} \\ F_{0y} \end{array} \right\} + [A] \left\{ \begin{array}{c} x(t)/R \\ y(t)/R \end{array} \right\} \qquad (10.3)$$

$$\left\{ \begin{array}{c} M_x(t) \\ M_y(t) \end{array} \right\} = \left\{ \begin{array}{c} M_{0x} \\ M_{0y} \end{array} \right\} + [B] \left\{ \begin{array}{c} x(t)/R \\ y(t)/R \end{array} \right\} \qquad (10.4)$$

The magnitude of the dimensionless radial force will be denoted by $F_0 = (F_{0x}^2 + F_{0y}^2)^{\frac{1}{2}}$, and its direction, θ, will be measured from the tongue or cutwater of the volute in the direction of rotation.

One particular feature of the rotordynamic matrices, $[A]$ and $[B]$, deserves special note. There are many geometries in which the rotordynamic forces should be invariant to a rotation of the x, y axes. Such will be the case only if

$$A_{xx} = A_{yy}; \quad A_{xy} = -A_{yx} \tag{10.5}$$

$$B_{xx} = B_{yy}; \quad B_{xy} = -B_{yx} \tag{10.6}$$

This does appear to be the case for virtually all of the experimental measurements that have been made in turbomachines.

The prototypical displacement will clearly consist of a circular whirl motion of "eccentricity", ϵ, and whirl frequency, ω, so that $x(t) = \epsilon \cos \omega t$ and $y(t) = \epsilon \sin \omega t$. As indicated in figure 10.1, an alternative notation is to define "rotordynamic forces", F_n^* and F_t^*, that are normal and tangential to the circular whirl orbit at the instantaneous position of the center of rotation. Note that F_n^* is defined as positive outward and F_t^* as positive in the direction of rotation, Ω. It follows that

$$F_n^* = \epsilon \left(A_{xx}^* + A_{yy}^* \right) / 2 \tag{10.7}$$

$$F_t^* = \epsilon \left(A_{yx}^* - A_{xy}^* \right) / 2 \tag{10.8}$$

and it is appropriate to define dimensionless normal and tangential forces, F_n and F_t, by dividing by $\rho \pi \Omega^2 R^2 L \epsilon$. Then the conditions of rotational invariance can be restated as

$$A_{xx} = A_{yy} = F_n \tag{10.9}$$

$$A_{yx} = -A_{xy} = F_t \tag{10.10}$$

Since this condition is met in most of the experimental data, it becomes convenient to display the rotordynamic forces by plotting F_n and F_t as functions of the geometry, operating condition and frequency ratio, ω / Ω. This presentation of the rotordynamic forces has a number of advantages from the perspective of physical interpretation. In many applications the normal force, F_n, is modest compared with the potential restoring forces which can be generated by the bearings and the casing. The tangential force has greater significance for the stability of the rotor system. Clearly a tangential force that is in the same direction as the whirl velocity ($F_t > 0$ for $\omega > 0$ or $F_t < 0$ for $\omega < 0$) will be rotordynamically destabilizing, and will cause a fluid-induced reduction in the critical whirl speeds of the machine. On the other hand, an F_t in the opposite direction to ω will be whirl stabilizing.

Furthermore, it is conventional among rotordynamicists to decompose the matrix $[A]$ into added mass, damping and stiffness matrices according to

$$[A] \begin{Bmatrix} x/R \\ y/R \end{Bmatrix} = - \begin{bmatrix} M & m \\ -m & M \end{bmatrix} \begin{Bmatrix} \ddot{x}/R\Omega^2 \\ \ddot{y}/R\Omega^2 \end{Bmatrix} - \begin{bmatrix} C & c \\ -c & C \end{bmatrix} \begin{Bmatrix} \dot{x}/R\Omega \\ \dot{y}/R\Omega \end{Bmatrix} - \begin{bmatrix} K & k \\ -k & K \end{bmatrix} \begin{Bmatrix} x/R \\ y/R \end{Bmatrix} \tag{10.11}$$

where the dot denotes differentiation with respect to time, so that the added mass matrix, $[M]$, multiplies the acceleration vector, the damping matrix, $[C]$, multiplies the velocity vector, and the stiffness matrix, $[K]$, multiplies the displacement vector. Note that the above has assumed rotational invariance of $[A]$, $[M]$, $[C]$ and $[K]$; M and m are respectively termed the direct and cross-coupled added mass, C and c the direct and cross-coupled damping, and K and k the direct and cross-coupled stiffness. Note also that the corresponding dimensional rotordynamic coefficients, M^*, m^*, C^*, c^*, K^*, and k^* are related to the dimensionless versions by

$$M, m = \frac{M^*, m^*}{\rho \pi R^2 L}; \quad C, c = \frac{C^*, c^*}{\rho \pi R^2 L \Omega}; \quad K, k = \frac{K^*, k^*}{\rho \pi R^2 L \Omega^2} \tag{10.12}$$

The representation of equation 10.11 is equivalent to assuming a quadratic dependence of the elements of $[A]$ (and the forces F_n, F_t) on the whirl frequency, or frequency ratio, ω/Ω. It should be emphasized that fluid mechanical forces do not always conform to such a simple frequency dependence. For example, in section 10.6, we shall encounter a force proportional to $\omega^{\frac{3}{2}}$. Nevertheless, it is of value to the rotordynamicists to fit quadratics to the plots of F_n and F_t against ω/Ω, since, from the above relations, it follows that

$$F_n = M (\omega/\Omega)^2 - c (\omega/\Omega) - K \tag{10.13}$$

$$F_t = -m (\omega/\Omega)^2 - C (\omega/\Omega) + k \tag{10.14}$$

and, therefore, all six rotordynamic coefficients can be directly evaluated from quadratic curve fits to the graphs of F_n and F_t against ω/Ω.

Since m is often small and is frequently assumed to be negligible, the sign of the tangential force is approximately determined by the quantity $k\Omega/\omega C$. Thus rotordynamicists often seek to examine the quantity $k/C = k^*/\Omega C^*$, which is often called the "whirl ratio" (not to be confused with the whirl frequency ratio, ω/Ω). Clearly larger values of this whirl ratio imply a larger range of frequencies for which the tangential force is destabilizing and a greater chance of rotordynamic instability.

In the last few paragraphs we have focused on the forces, but it is clear that a parallel construct is relevant to the rotordynamic moments. It should be recognized that each of the components of a turbomachine will manifest its own rotordynamic coefficients which will all need to be included in order to effect a complete rotordynamic analysis of the machine. The methods used in such rotordynamic analyses are beyond the scope of this book. However, we shall attempt to review the origin of these forces in the bearings, seals, and other components of the turbomachine. Moreover, both the main flow and leakage flows associated with the impeller will generate contributions. In order to permit ease of comparison between the rotordynamic effects contributed by the various components, we shall use a similar nondimensionalization for all the components.

10.3 Hydrodynamic Bearings and Seals

Hydrodynamic bearings, seals, and squeeze-film dampers constitute a class of devices that involve the flow in an annulus between two cylinders; the inner cylinder is generally the shaft (radius, R) which is rotating at a frequency, Ω, and may also be whirling with an amplitude or eccentricity, ϵ, and a frequency, ω. The outer cylinder is generally static and fixed to the support structure. The mean clearance (width of the annulus) will be denoted by δ, and the axial length by L. In both hydrodynamic bearings and seals, the basic fluid motion is caused by the rotation of the shaft. In a seal, there is an additional axial flow due to the imposed axial pressure difference. In a squeeze-film damper, there is no rotational motion, but forces are generated by the whirl motion of the "rotor."

The Reynolds number is an important parameter in these flows, and it is useful to evaluate three different Reynolds numbers based on the rotational velocity, on the mean axial velocity, V (given by $V = Q/2\pi R\delta$ where Q is the volumetric axial flow rate), and on the velocity associated with the whirl motion. These are termed the rotational, axial and whirl Reynolds numbers and are defined, respectively, by

$$Re_\Omega = \Omega R\delta/v, \quad Re_V = V\delta/v, \quad Re_\omega = \omega R\delta/v \qquad (10.15)$$

where v is the kinematic viscosity of the fluid in the annulus. In a hydrodynamic bearing, the fluid must be of sufficiently high viscosity so that $Re_\Omega \ll 1$. This is because the bearing depends for its operation on a large fluid restoring force or stiffness occurring when the shaft or rotor is displaced from a concentric position. Typically a bearing will run with a mean eccentricity that produces the fluid forces that counteract the rotor weight or other radial forces. It is important to recognize that the fluid only yields such a restoring force or stiffness when the flow in the annulus is dominated by viscous effects. For this to be the case, it is necessary that $Re_\Omega \ll 1$. If this is not the case, and $Re_\Omega \gg 1$ then, as we shall discuss later, the sign of the fluid force is reversed, and, instead of tending to decrease eccentricity, the fluid force tends to magnify it. This is called the "Bernoulli effect" or "inertia effect", and can be simply explained as follows. When an eccentricity is introduced, the fluid velocities will be increased over that part of the rotor circumference where the clearance has been reduced. At Reynolds numbers much larger than unity, the Bernoulli equation is applicable, and higher velocities imply lower pressure. Therefore the pressure in the fluid is decreased where the clearance is small and, consequently, there will be a net force on the rotor in the direction of the displacement. This "negative stiffness" ($K < 0$) is important in the rotordynamics of seals and impellers.

Another parameter of importance is the ratio of the axial length to radius, L/R, of the bearing or seal. For large L/R, the predominant fluid motions caused by the rotordynamic perturbations occur in the circumferential direction. On the other hand, in a short seal or bearing, the predominant effect of the rotordynamic perturbation is

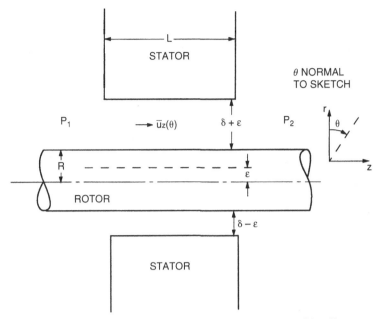

Figure 10.2. Schematic of a short seal demonstrating the Lomakin effect.

to cause circumferential variation in the axial fluid velocity. This gives rise to the so-called "Lomakin effect" in short seals operating at high Reynolds numbers (Lomakin 1958). The circumstances are sketched in figure 10.2, in which we use a cylindrical coordinate system, (r, θ, z), to depict a plain annular seal with a clearance, δ. The fluid velocity, u_z, is caused by the pressure difference, $\Delta p = (p_1 - p_2)$. We denote the axial velocity averaged over the clearance by $\bar{u}_z$, and this will be a function of θ when the rotor is displaced by an eccentricity, ϵ. The Lomakin effect is caused by circumferential variations in the entrance losses in this flow. On the side with the smaller clearance, the entrance losses are smaller because $\bar{u}_z$ is smaller. Consequently, the mean pressure is larger on the side with the smaller clearance, and the result is a restoring force due to this circumferential pressure distribution. This is known as the Lomakin effect, and gives rise to a positive fluid-induced stiffness, K. Note that the competing Bernoulli and Lomakin effects can cause the sign of the fluid-induced stiffness of a seal to change as the geometry changes.

In the following sections we examine more closely some of the fluid-induced rotordynamic effects in bearings, seals, and impellers.

10.4 Bearings at Low Reynolds Numbers

The rotordynamics of a simple hydrodynamic bearing operating at low Reynolds number ($Re_\Omega \ll 1$) will be examined first. The conventional approach to this problem (Pinkus and Sternlicht 1961) is to use Reynolds' approximate equation for the fluid motions in a thin film. In the present context, in which the fluid is contained between

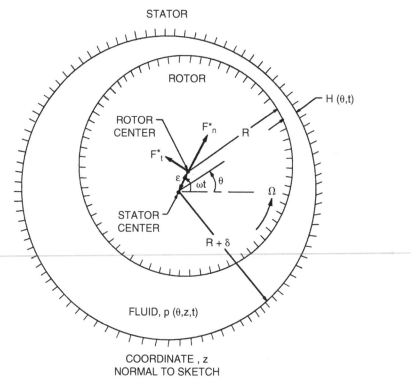

Figure 10.3. Schematic of fluid-filled annulus between a stator and a rotating and whirling rotor.

two circular cylinders (figure 10.3) this equation becomes

$$\frac{1}{R^2}\frac{\partial}{\partial\theta}\left(H^3\frac{\partial p}{\partial\theta}\right) + \frac{\partial}{\partial z}\left(H^3\frac{\partial p}{\partial z}\right) = 6\mu\left\{2\frac{\partial H}{\partial t} + \frac{1}{R}\frac{\partial}{\partial\theta}(HU)\right\} \tag{10.16}$$

where (θ,z) are the circumferential and axial coordinates. This equation must be solved to find the pressure, $p(\theta,z,t)$, in the fluid (averaged over the radial extent of the clearance gap) given the clearance, $H(\theta,t)$, and the surface velocity, U, of the inner cylinder ($U = \Omega R$). An eccentricity, ϵ, at a whirl frequency of ω leads to a clearance, H, given by $\delta - \epsilon\cos(\omega t - \theta)$ and substituting for H completes the formulation of equation 10.16 for the pressure.

The rotordynamic forces, F_n^* and F_t^*, then follow from

$$\begin{Bmatrix} F_n^* \\ F_t^* \end{Bmatrix} = R \int_0^L \int_0^{2\pi} p \begin{Bmatrix} -\cos(\omega t - \theta) \\ \sin(\omega t - \theta) \end{Bmatrix} d\theta dz \tag{10.17}$$

where L is the axial length of the bearing.

Two simple asymptotic solutions are readily forthcoming for linear perturbations in which $\epsilon \ll \delta$. The first is termed the "long bearing" solution, and assumes, as discussed in the last section, that the dominant perturbations to the velocity occur in the circumferential velocities rather than the axial velocities. It follows that the second term in equation 10.16 can be neglected as small relative to the first term. Neglecting, in addition, all terms quadratic or higher order in ϵ, integration of equation 10.16 leads to

$$p = \frac{6\mu R^2 \epsilon}{\delta^3}(\Omega - 2\omega)\sin(\omega t - \theta) + \frac{12\mu R^2}{\delta^3}\frac{d\epsilon}{dt}\cos(\omega t - \theta) \qquad (10.18)$$

and to the following rotordynamic forces:

$$F_n^* = -\frac{12\pi\mu R^3 L}{\delta^3}\frac{d\epsilon}{dt} \; ; \; F_t^* = \frac{6\pi\mu R^3 L\epsilon}{\delta^3}(\Omega - 2\omega) \qquad (10.19)$$

In steady whirling motion, $d\epsilon/dt = 0$. The expression for F_t^* implies the following rotordynamic coefficients:

$$C^* = \frac{2k^*}{\Omega} = \frac{12\pi\mu R^3 L}{\delta^3} \qquad (10.20)$$

and $K^* = c^* = M^* = m^* = 0$.

The second, or "short bearing", solution assumes that the dominant perturbations to the velocities occur in the axial velocities; this usually requires L/R to be less than about 0.5. Then, assuming that the pressure is measured relative to a uniform and common pressure at both ends, $z = 0$ and $z = L$, integration of equation 10.16 leads to

$$p = z(L - z)\left[\frac{6\mu}{\delta^3}\frac{d\epsilon}{dt}\cos(\omega t - \theta) - \frac{3\mu\epsilon}{\delta^3}(\Omega - 2\omega)\sin(\omega t - \theta)\right] \qquad (10.21)$$

and, consequently,

$$F_n^* = -\frac{\pi\mu RL^3}{\delta^3}\frac{d\epsilon}{dt} \; ; \; F_t^* = \frac{\pi\mu RL^3\epsilon}{2\delta^3}(\Omega - 2\omega) \qquad (10.22)$$

Therefore, in the short bearing case,

$$C^* = \frac{2k^*}{\Omega} = \frac{\pi\mu RL^3}{2\delta^3} \qquad (10.23)$$

in contrast to the result in equation 10.20. Notice that, for both the long and short bearing, the value of the whirl ratio, $k^*/\Omega C^*$, is 0.5. Later, we will compare this value with that obtained for other flows and other devices.

It is particularly important to note that the tangential forces in both the long and short bearing solutions are negative for $\Omega < 2\omega$, and become positive for $\Omega > 2\omega$. This explains the phenomenon of "oil whip" in hydrodynamic bearings, first described by

Newkirk and Taylor (1925). They reported that violent shaft motions occurred when the shaft speed reached a value twice the critical speed of the shaft. This phenomenon is the response of a dynamic system at its natural frequency when the exciting tangential force becomes positive, namely when $\Omega > 2\omega$ (see Hori 1959). It is of interest to note that a similar critical condition occurs for high Reynolds number flow in the film (see equations 10.36 and 10.37).

The simple linear results described above can be augmented in several ways. First, similar solutions can be generated for the more general case in which the eccentricity is not necessarily small compared with the clearance. The results (Vance 1988) for the long bearing become

$$F_n^* = -\frac{12\pi\mu R^3 L}{\delta^3}\frac{1}{(1-\epsilon^2/\delta^2)^{\frac{3}{2}}}\frac{d\epsilon}{dt} \tag{10.24}$$

$$F_t^* = \frac{6\pi\mu R^3 L(\Omega-2\omega)}{\delta^3}\frac{\epsilon}{(1-\epsilon^2/\delta^2)(2+\epsilon^2/\delta^2)^{\frac{3}{2}}} \tag{10.25}$$

and, in the short bearing case,

$$F_n^* = -\frac{\pi\mu RL^3}{\delta^3}\frac{(1+2\epsilon^2/\delta^2)}{(1-\epsilon^2/\delta^2)^{\frac{5}{2}}}\frac{d\epsilon}{dt} \tag{10.26}$$

$$F_t^* = \frac{\pi\mu RL^3(\Omega-2\omega)}{2\delta^3}\frac{\epsilon}{(1-\epsilon^2/\delta^2)^{\frac{3}{2}}} \tag{10.27}$$

These represent perhaps the only cases in which rotordynamic forces and coefficients can be evaluated for values of the eccentricity, ϵ, comparable with the clearance, δ. The nonlinear analysis leads to rotordynamic coefficients which are functions of the eccentricity, ϵ, and the variation with ϵ/δ is presented graphically in figure 10.4. Note that the linear values given by equations 10.20 and 10.23 are satisfactory up to ϵ/δ of the order of 0.5.

Second, it is important to note that cavitation or gas dissolution in liquid-filled bearings can often result in a substantial fraction of the annulus being filled by a gas bubble or bubbles. The reader is referred to Dowson and Taylor (1979) for a review of this complicated subject. Quite crude approximations are often introduced into lubrication analyses in order to try to account for this "cavitation." The most common approximation is to assume that the two quadrants in which the pressure falls below the mean are completely filled with gas (or vapor) rather than liquid. Called a π-film cavitated bearing, this heuristic assumption leads to the following rotordynamic forces

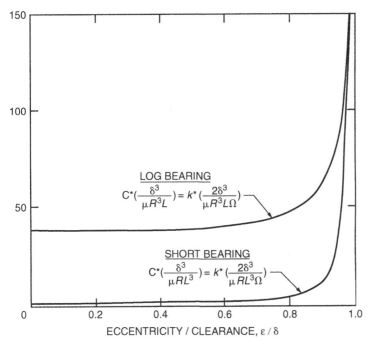

Figure 10.4. Dimensionless damping and cross-coupled stiffness for noncavitating long and short bearings as functions of the eccentricity ratio, ϵ/δ.

(Vance 1988). For the cavitated (π-film) long bearing

$$F_n^* = -\frac{6\mu R^3 L}{\delta^2}\left[\frac{2|\Omega - 2\omega|\epsilon^2}{\delta^2(2+\epsilon^2/\delta^2)(1-\epsilon^2/\delta^2)} + \frac{\pi d\epsilon/dt}{\delta(1-\epsilon^2/\delta^2)^{\frac{3}{2}}}\right] \quad (10.28)$$

$$F_t^* = \frac{6\mu R^3 L}{\delta^3}\left[\frac{(\Omega - 2\omega)\pi\epsilon}{(2+\epsilon^2/\delta^2)(1-\epsilon^2/\delta^2)^{\frac{3}{2}}} + \frac{4d\epsilon/dt}{(1+\epsilon/\delta)(1-\epsilon^2/\delta^2)}\right] + 2RLp_0 \quad (10.29)$$

where p_0 is the pressure in the cavity. These expressions are similar to, but not identical with, the expressions derived by Hori (1959) and used to explain oil whip. For the cavitated (π-film) short bearing

$$F_n^* = -\frac{\mu R L^3}{\delta^2}\left[\frac{|\Omega - 2\omega|\epsilon^2}{\delta^2(1-\epsilon^2/\delta^2)^2} + \frac{\pi(1+2\epsilon^2/\delta^2)d\epsilon/dt}{2\delta(1-\epsilon^2/\delta^2)^{\frac{5}{2}}}\right] \quad (10.30)$$

$$F_t^* = \frac{\mu R L^3}{\delta^2}\left[\frac{(\Omega - 2\omega)\pi\epsilon}{4\delta(1-\epsilon^2/\delta^2)^{\frac{3}{2}}} + \frac{2\epsilon d\epsilon/dt}{\delta^2(1-\epsilon^2/\delta^2)^2}\right] + 2RLp_0 \quad (10.31)$$

It would, however, be appropriate to observe that rotordynamic coefficients under cavitating conditions remain to be measured experimentally, and until such tests are performed the above results should be regarded with some scepticism.

Finally, we note that all of the fluid inertial effects have been neglected in the above analyses, and, consequently, the question arises as to how the results might change when the Reynolds number, Re_Ω, is no longer negligibly small. Such analyses require a return to the full Navier-Stokes equations, and the author has explored the solutions of these equations in the case of long bearings (Brennen 1976). In the case of whirl with constant eccentricity ($d\epsilon/dt = 0$), it was shown that there are two separate sets of asymptotic results for $Re_\omega \ll \delta^3/R^3$, and for $\delta^3/R^3 \ll Re_\omega \ll \delta^2/R^2$. For $Re_\omega \ll \delta^3/R^3$, the rotordynamic forces are

$$F_n = -\frac{9}{4}\frac{R^5}{\delta^5}\left(1 - \frac{2\omega}{\Omega}\right) \tag{10.32}$$

while F_t^* is the same as given in equation 10.19. Notice that equation 10.32 implies a direct stiffness, K, and cross-coupled damping, c, given by

$$K = \frac{c}{2} = \frac{9}{4}\frac{R^5}{\delta^5} \tag{10.33}$$

On the other hand, for the range of Reynolds numbers given by $\delta^3/R^3 \ll Re_\omega \ll \delta^2/R^2$, the rotordynamic forces are

$$F_n = \frac{16\delta}{R\,Re_\Omega^2}\left(\frac{2\omega}{\Omega} - 1\right); \quad F_t = \frac{128\delta^4}{3R^4 Re_\Omega^3}\left(1 - \frac{2\omega}{\Omega}\right) \tag{10.34}$$

so that

$$K = \frac{c}{2} = \frac{16\delta}{R\,Re_\Omega^2}; \quad C = 2k = \frac{256\delta^4}{3R^4 Re_\Omega^3} \tag{10.35}$$

In both cases the direct stiffness, K, is positive, implying a positive hydrodynamic restoring force caused by the inertial terms in the equations of fluid motion.

10.5 Annulus at High Reynolds Numbers

Consider now the flows of the last section when the Reynolds numbers become much greater than unity. The name "bearing" must be omitted, since the flow no longer has the necessary rotordynamic characteristics to act as a hydrodynamic bearing. Nevertheless, such flows are of interest since there are many instances in which rotors are surrounded by fluid annuli. Fritz (1970) used an extension of a lubrication theory in which he included fluid inertia and fluid frictional effects for several types of flow in the annulus, including Taylor vortex flow and fully turbulent flow. Though some of his arguments are heuristic, the results are included here because of their practical value. The rotordynamic forces which he obtains are

$$F_n^* = \frac{\pi \rho R^3 L}{\delta} \left[\epsilon \left(\frac{\Omega}{2} - \omega \right)^2 - \frac{d^2 \epsilon}{dt^2} - \Omega f \frac{d\epsilon}{dt} \right] \tag{10.36}$$

$$F_t^* = \frac{\pi \rho R^3 L}{\delta} \left[\Omega \epsilon f \left(\frac{\Omega}{2} - \omega \right) + (\Omega - 2\omega) \frac{d\epsilon}{dt} \right] \tag{10.37}$$

where f is a fluid friction term that varies according to the type of flow in the annulus. For laminar flow, $f = 12\nu/\delta^2 \Omega$ and the first term in the square bracket of F_t^* and the last term in F_n^* are identical to the forces for a noncavitating long bearing as given in equation 10.19. But, Fritz also constructs forms for f for Taylor vortex flow and for turbulent flow. For example, for turbulent flow

$$f = 1.14 f_T R/\delta \tag{10.38}$$

where f_T is a friction factor that correlates with the Reynolds number, Re_Ω.

The other terms in equations 10.36 and 10.37 that do not involve f are caused by the fluid inertia and are governed by the added mass, $M^* = \pi \rho R^3 L/\delta$, which Fritz confirms by experimental measurements. Note that equations 10.36 and 10.37 imply rotordynamic coefficients as follows:

$$M = R/\delta; \quad c = R/\delta; \quad K = -R/4\delta$$
$$m = 0; \quad C = f R/\delta; \quad k = f R/2\delta \tag{10.39}$$

The author also examined these flows using solutions to the Navier-Stokes equations (Brennen 1976). For annuli in which δ is not necessarily small compared with R, the added mass becomes

$$M^* = \frac{\pi \rho L R^2 (R_S^2 + R^2)}{(R_S^2 - R^2)} \tag{10.40}$$

where R_S is the radius of the rigid stator.

10.6 Squeeze Film Dampers

A squeeze film damper consists of a nonrotating cylinder surrounded by a fluid annulus contained by an outer cylinder. A shaft runs within the inner, nonrotating cylinder so that the latter may perform whirl motions without rotation. The fluid annulus is intended to damp any rotordynamic motions of the shaft. It follows that figure 10.3 can also represent a squeeze film damper as long as Ω is set to zero. The device is intended to operate at low Reynolds numbers, Re_ω, and several of the results already described can be readily adopted for use in a squeeze-film damper. Clearly analyses can be generated for both long and short squeeze film dampers. The long squeeze film damper is one flow for which approximate solutions to the full Navier-Stokes equations can be found (Brennen 1976). Two sets of asymptotic results emerge, depending on whether Re_ω is much less than, or much greater than, $72R/\delta$. In the case of thin films ($\delta \ll R$), the rotordynamic forces for $Re_\omega \ll 72R/\delta$ are

$$F_n^* = 6\pi\rho R^3 L\omega^2\epsilon/5\delta \tag{10.41}$$

$$F_t^* = 12\pi\mu R^3 L\omega\epsilon/\delta^3 \tag{10.42}$$

where the F_t^* is the same as that for a noncavitating long bearing. On the other hand, for $Re_\omega \gg 72R/\delta$:

$$F_n^* = \frac{\pi\rho R^3 L\omega^2\epsilon}{\delta}\left[1+\left(\frac{2\nu}{\omega\delta^2}\right)^{\frac{1}{2}}\right] \tag{10.43}$$

$$F_t^* = \pi\rho R^2 L\omega^2\epsilon\left(\frac{2\nu}{\omega\delta^2}\right)^{\frac{1}{2}} \tag{10.44}$$

The $\omega^{\frac{3}{2}}$ dependent terms in these relations are very unfamiliar to rotordynamicists. However, such frequency dependence is common in flows that are dominated by the diffusion of vorticity.

The relations 10.41 to 10.44 are limited to small amplitudes, $\epsilon \ll \delta$, and to values of $\omega\epsilon^2/\nu \ll 1$. At larger amplitudes and Reynolds numbers, $\omega\epsilon^2/\nu$, it is necessary to resort to lubrication analyses supplemented, where necessary, with inertial terms in the same manner as described in the last section. Vance (1988) delineates such an approach to squeeze film dampers.

10.7 Turbulent Annular Seals

In an annular seal, the flows are usually turbulent because of the high Reynolds numbers at which they operate. In this section we describe the approaches taken to identify the rotordynamic properties of these flows. Black and his co-workers (Black 1969, Black and Jensen 1970) were the first to attempt to identify and model the rotordynamics of turbulent annular seals. Bulk flow models (similar to those of Reynolds lubrication equations) were used. These employ velocity components, $\bar{u}_z(z,\theta)$ and $\bar{u}_\theta(z,\theta)$, that are averaged over the clearance. Black and Jensen used several heuristic assumptions in their model, such as the assumption that $\bar{u}_\theta = R\Omega/2$. Moreover, their governing equations do not reduce to recognizable turbulent lubrication equations. These issues caused Childs (1983b) to publish a revised version of the bulk flow model and we will focus on Childs' model here. Childs (1987, 1989) has also employed a geometric generalization of the same bulk flow model to examine the rotordynamic characteristics of discharge-to-suction leakage flows around shrouded

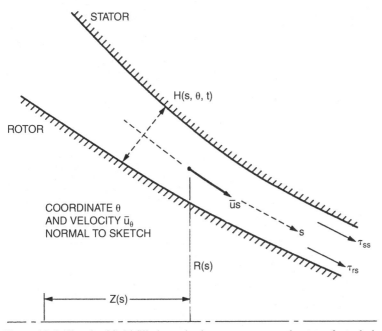

Figure 10.5. Sketch of fluid filled annulus between a rotor and a stator for turbulent lubrication analysis.

centrifugal pump impellers, and it is therefore convenient to include here the more general form of his analysis. The geometry is sketched in figure 10.5, and is described by coordinates of the meridian of the gap as given by $Z(s)$ and $R(s)$, $0 < s < L$, where the coordinate, s, is measured along that meridian. The clearance is denoted by $H(s,\theta,t)$ where the unperturbed value of H is $\delta(s)$. The equations governing the bulk flow are averaged over the clearance. This leads to a continuity equation of the form

$$\frac{\partial H}{\partial t} + \frac{\partial}{\partial s}(H\bar{u}_s) + \frac{1}{R}\frac{\partial}{\partial \theta}(H\bar{u}_\theta) + \frac{H}{R}\frac{dR}{ds}\bar{u}_s = 0 \qquad (10.45)$$

where $\bar{u}_s$ and $\bar{u}_\theta$ are velocities averaged over the local clearance. The meridional and circumferential momentum equations are

$$-\frac{1}{\rho}\frac{\partial p}{\partial s} = \frac{\tau_{ss}}{\rho H} + \frac{\tau_{sr}}{\rho H} - \frac{\bar{u}_\theta^2}{R}\frac{dR}{ds} + \frac{\partial \bar{u}_s}{\partial t} + \frac{\bar{u}_\theta}{R}\frac{\partial \bar{u}_s}{\partial \theta} + \bar{u}_s\frac{\partial \bar{u}_s}{\partial \theta} \qquad (10.46)$$

$$-\frac{1}{\rho R}\frac{\partial p}{\partial \theta} = \frac{\tau_{\theta s}}{\rho H} + \frac{\tau_{\theta r}}{\rho H} + \frac{\partial \bar{u}_\theta}{\partial t} + \frac{\bar{u}_\theta}{R}\frac{\partial \bar{u}_\theta}{\partial \theta} + \bar{u}_s\frac{\partial \bar{u}_\theta}{\partial s} + \frac{\bar{u}_\theta \bar{u}_r}{R}\frac{dR}{ds} \qquad (10.47)$$

The approach used by Hirs (1973) is employed to determine the turbulent shear stresses, τ_{ss} and $\tau_{\theta s}$, applied to the stator by the fluid in the s and θ directions respectively:

$$\frac{\tau_{ss}}{\rho \bar{u}_s} = \frac{\tau_{\theta s}}{\rho \bar{u}_\theta} = \frac{A_s \bar{u}_s}{2}\left[1 + (\bar{u}_\theta/\bar{u}_s)^2\right]^{\frac{m_s+1}{2}}(Re_s)^{m_s} \qquad (10.48)$$

and the stresses, τ_{sr} and $\tau_{\theta r}$, applied to the rotor by the fluid in the same directions:

$$\frac{\tau_{sr}}{\rho \bar{u}_s} = \frac{\tau_{\theta r}}{\rho(\bar{u}_\theta - \Omega R)} = \frac{A_r \bar{u}_s}{2} \left[1 + \left\{ (\bar{u}_\theta - \Omega R)/\bar{u}_s \right\}^2 \right]^{\frac{m_\theta + 1}{2}} (Re_s)^{m_\theta} \qquad (10.49)$$

where the local meridional Reynolds number

$$Re_s = H\bar{u}_s / \nu \qquad (10.50)$$

and the constants A_s, A_r, m_s, and m_θ are chosen to fit the available data on turbulent shear stresses. Childs (1983a) uses typical values of these constants

$$A_s = A_r = 0.0664; \quad m_s = m_\theta = -\frac{1}{4} \qquad (10.51)$$

The clearance, pressure, and velocities are divided into mean components (subscript 0) that would pertain in the absence of whirl, and small, linear perturbations (subscript 1) due to an eccentricity, ϵ, rotating at the whirl frequency, ω:

$$H(s,\theta,t) = H_0(s) + \epsilon Re\left\{ H_1(s)e^{i(\theta - \omega t)} \right\}$$

$$p(s,\theta,t) = p_0(s) + \epsilon Re\left\{ p_1(s)e^{i(\theta - \omega t)} \right\}$$

$$\bar{u}_s(s,\theta,t) = \bar{u}_{s0}(s) + \epsilon Re\left\{ \bar{u}_{s1}(s)e^{i(\theta - \omega t)} \right\}$$

$$\bar{u}_\theta(s,\theta,t) = \bar{u}_{\theta 0}(s) + \epsilon Re\left\{ \bar{u}_{\theta 1}(s)e^{i(\theta - \omega t)} \right\} \qquad (10.52)$$

These expressions are substituted into the governing equations listed above to yield a set of equations for the mean flow quantities (p_0, $\bar{u}_{s0}$, and $\bar{u}_{\theta 0}$), and a second set of equations for the perturbation quantities (p_1, $\bar{u}_{s1}$, and $\bar{u}_{\theta 1}$); terms which are of quadratic or higher order in ϵ are neglected.

With the kind of complex geometry associated, say, with discharge-to-suction leakage flows in centrifugal pumps, it is necessary to solve both sets of equations numerically in order to evaluate the pressures, and then the forces, on the rotor. However, with the simple geometry of a plain, untapered annular seal where

$$R(s) = R, \quad H_0(s) = \delta, \quad s = z, \quad H_1(s) = 1 \qquad (10.53)$$

and in which

$$\bar{u}_{s0} = \frac{Q}{2\pi R\delta} = V \qquad (10.54)$$

where Q is the volumetric flow rate, Childs (1983a) was able to obtain analytic solutions to both the mean and perturbation equations. The resulting evaluation of the rotordynamic forces leads to the following rotordynamic coefficients:

$$K = \left(\frac{2\Delta p^T}{\rho V^2}\right)\phi^2 \frac{R}{2\lambda_1 L}\left[\mu_0 - \mu_2(L/2\phi R)^2\right] \tag{10.55}$$

$$C = 2k = \left(\frac{2\Delta p^T}{\rho V^2}\right)\frac{\phi\mu_1}{2\lambda_1} \tag{10.56}$$

$$M = c = \left(\frac{2\Delta p^T}{\rho V^2}\right)\frac{\mu_2 L}{2\lambda_1 R} \tag{10.57}$$

where ϕ is the flow coefficient ($\phi = V/\Omega R$), and Δp^T is the total pressure drop across the seal where

$$\frac{2\Delta p^T}{\rho V^2} = 1 + C_E L + 2\lambda_2 \tag{10.58}$$

and λ, μ_0, μ_1, and μ_2 are given by

$$\lambda_1 = 0.0664(Re_V)^{-\frac{1}{4}}\left\{1 + 1/4\phi^2\right\}^{\frac{3}{8}} \tag{10.59}$$

$$\lambda_2 = \lambda_1 L/\delta \tag{10.60}$$

$$\mu_0 = 5\lambda_2^2\mu_5/2(1 + C_{EL} + 2\lambda_2) \tag{10.61}$$

$$\mu_1 = 2\lambda_2\left\{\mu_5 + \frac{1}{2}\lambda_2\mu_4(\mu_5 + 1/6)\right\}/(1 + C_{EL} + 2\lambda_2) \tag{10.62}$$

$$\mu_2 = \lambda_2(\mu_5 + 1/6)/(1 + C_{EL} + 2\lambda_2) \tag{10.63}$$

$$\mu_4 = (1 + 7\phi^2)/(1 + 4\phi^2) \tag{10.64}$$

$$\mu_5 = (1 + C_{EL})/2(1 + C_{EL} + \mu_4\lambda_2) \tag{10.65}$$

where C_{EL} is an entrance loss coefficient for which the data of Yamada (1962) was used. Note that there are two terms in K; the first, which contains μ_0, results from the Lomakin effect, while the second, involving μ_2, results from the Bernoulli effect (section 10.3).

The results obtained by Black and Jensen (1970) are similar to the above except for the expressions for some of the λ and μ quantities. Childs (1983a) contrasts the two sets of expressions, and observes that one of the primary discrepancies is that the Black and Jensen expressions yield a significant smaller added mass, M. We should also note that Childs (1983a) includes the effect of inlet preswirl which has a significant influence on the rotordynamic coefficients. Preswirl was not included in the results presented above.

Typical results from the expressions 10.55 to 10.57 are presented in figures 10.6 and 10.7, which show the variations with flow coefficient, ϕ, and the geometric

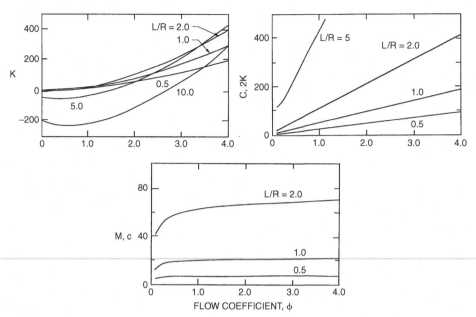

Figure 10.6. Typical dimensionless rotordynamic coefficients from Childs' (1983a) analysis of a plain, untapered and smooth annular seal with $\delta/R = 0.01$, $Re_V = 5000$, and $C_{EL} = 0.1$.

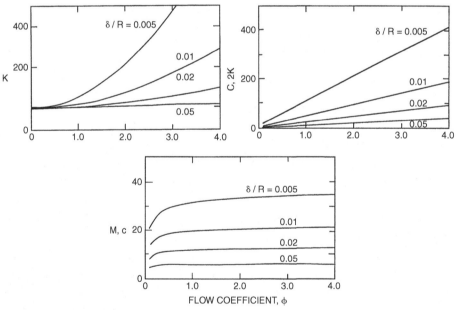

Figure 10.7. Typical dimensionless rotordynamic coefficients from Childs' (1983a) analysis of a plain, untapered and smooth annular seal with $L/R = 1$, $Re_V = 5000$, and $C_{EL} = 0.1$.

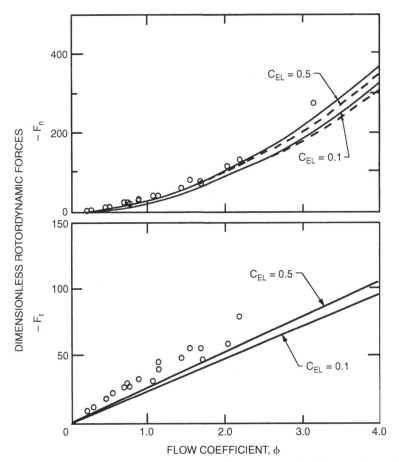

Figure 10.8. The measurements by Childs and Dressman (1982) of the rotordynamic forces for a straight, smooth annular seal ($L/R = 1.0, \delta/R = 0.01$) for a range of Reynolds numbers, $2205 < Re_V < 13390$, and under synchronous excitation. Also shown are the predictions of the theory of Childs (1983a) for $Re_V = 10000$ (solid lines) and 15000 (dashed lines) and two different entrance loss coefficients, C_{EL}, as shown.

ratios, L/R and δ/R. The effects of Reynolds number, Re_V, and of the entrance loss coefficient, are small as demonstrated in figure 10.8. Note the changes in sign in the direct stiffness, K, that result from the Lomakin effect becoming larger than the Bernoulli effect, or vice-versa. Note, also, that the whirl ratio, k/C, is 0.5 in all cases.

Childs and Dressman (1982) have published experimental measurements of the rotordynamic forces in a plain, smooth, annular seal with a length, L, to radius, R, ratio of 1.0, a clearance, δ, to radius ratio of 0.01 at various flow rates and speeds. The excitation was synchronous ($\omega/\Omega = 1$) so that

$$F_n = M - c - K; \quad F_t = -m - C + k \tag{10.66}$$

Consequently, if one assumes the theoretical results $M = c$, $m = 0$ and $C = 2k$ to be correct, then

$$F_n = -K; \quad F_t = -k = -\frac{c}{2} \tag{10.67}$$

The data of Childs and Dressman for Reynolds numbers in the ranges $2205 < Re_V < 13390$ and $2700 < Re_\Omega < 10660$ are plotted in figure 10.8. It is readily seen that, apart from the geometric parameters L/R and δ/R, the rotordynamic characteristics are primarily a function of the flow coefficient, ϕ, defined as $\phi = V/\Omega R = Re_V/Re_\Omega$, and only depend weakly on the Reynolds number itself. The results from Childs' (1983a) theory using equations 10.55 to 10.57 are also shown and exhibit quite good agreement with the measurements. As can be seen, the theoretical results are also only weakly dependent on Re_V or the entrance loss coefficient, C_{EL}.

Nordmann and Massman (1984) conducted experiments on a similar plain annular seal with $L/R = 1.67$ and $\delta/R = 0.0167$, and measured the forces for both synchronous and nonsynchronous excitation. Thus, they were able to extract the rotordynamic coefficients M, C, c, K, and k. Their results for a Reynolds number, $Re_V = 5265$, are presented in figure 10.9, where they are compared with the corresponding predictions of Childs' (1983a) theory (using $C_{EL} = 0.1$). In comparing theory and experiment, we must remember that the results are quite insensitive to Reynolds number, and the theoretical data does not change much with changes in C_{EL}. Some of the Nordmann and Massmann data exhibits quite a lot of scatter; however, with the notable exception of the cross-coupled stiffness, k, the theory is in good agreement with the data. The reason for the discrepancy in the cross-coupled stiffness is unclear. However, one must bear in mind that the theory uses correlations developed from results for nominally *steady* turbulent flows, and must be regarded as tentative until there exists a greater understanding of unsteady turbulent flows.

In the last decade, a substantial body of data has been accumulated on the rotordynamic characteristics of annular seals, particularly as regards such geometric effects as taper, various kinds of roughness, and the effects of labyrinths. We include here only a few examples. Childs and Dressman (1985) conducted both theoretical and experimental investigations of the effect of taper on the synchronous rotordynamic forces. They showed that the introduction of a taper increases the leakage and the direct stiffness, K^*, but decreases the other rotordynamic coefficients. An optimum taper angle exists with respect to both the direct stiffness and the ratio of direct stiffness to leakage. Childs and Kim (1985) have examined the effects of directionally homogeneous surface roughness on both the rotor and the stator. Test results for four different surface roughnesses applied to the stator or casing (so-called "damper seals" that have smooth rotors) showed that the roughness increases the damping and decreases the leakage.

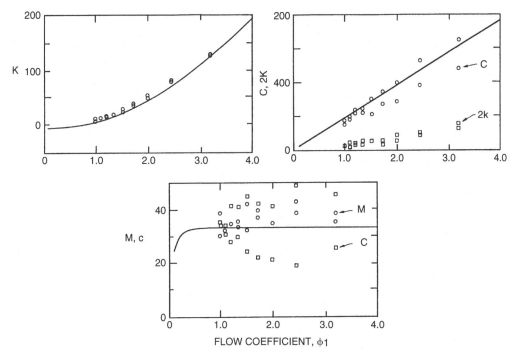

Figure 10.9. Dimensionless rotordynamic coefficients measured by Nordmann and Massmann (1984) for a plain seal with $L/R = 1.67$, $\delta/R = 0.0167$, and $Re_V = 5265$. Also shown are the corresponding theoretical results using Childs' (1983a) theory with $C_{EL} = 0.1$.

10.8 Labyrinth Seals

Labyrinth seals with teeth on either the rotor or the stator are frequently used, because the teeth help to minimize the leakage through the seals. However, the teeth also have rotordynamic consequences which have been explored by Wachter and Benckert (1980), Childs and Scharrer (1986), and others. Childs and Scharrer measured the stiffness and damping coefficients for some labyrinth seals, and reached the following conclusions. First, in all cases, the rotordynamic forces were independent of the rotational speed, Ω, and dependent on the axial pressure drop, Δp. The appropriate nondimensionalizing velocity is therefore the typical axial velocity caused by the axial pressure drop, $(2\Delta p/\rho)^{\frac{1}{2}}$. Childs and Scharrer suggest that the reason for this behavior is that the mean fluid motions are dominated by throughflow over and between the teeth, and that the shear caused by the rotation of the rotor has relatively little effect on the flow at the high Reynolds numbers involved.

Typical dimensionless values of the rotordynamic coefficients K, k, and C are presented in table 10.1, where we may observe that the cross-coupled stiffness, k, is smaller for the teeth-on-stator configuration. This means that, since the damping, C, is similar for the two cases, the teeth-on-stator configuration is more stable rotordynamically.

Table 10.1. *Rotordynamic characteristics of labyrinth seals with zero inlet swirl (data from Childs and Scharrer 1986).*

	Teeth on Rotor			Teeth on Stator		
	Mean	Min.	Max.	Mean	Min.	Max.
$K^*/2\pi \Delta pL$	-1.17	-1.03	-1.25	-0.62	-0.45	-0.74
$k^*/2\pi \Delta pL$	1.15	0.79	1.68	0.86	0.67	1.07
$C^*/\pi RL(2\rho\Delta p)^{\frac{1}{2}}$	0.0225	0.0168	0.0279	0.0219	0.0182	0.0244

However, Childs and Scharrer also found that the coefficients were very sensitive to the inlet swirl velocity upstream of the seal. In particular, the cross-coupled stiffness increased markedly with increased swirl in the same direction as shaft rotation. On the other hand, imposed swirl in a direction opposite to shaft rotation causes a reversal in the sign of the cross-coupled stiffness, and thus has a rotordynamically stabilizing effect.

10.9 Blade Tip Rotordynamic Effects

In a seminal paper, Alford (1965) identified several rotordynamic effects arising from the flow in the clearance region between the tip of an axial flow turbomachine blade and the static housing. However, the so-called "Alford effects" are only some of the members of a class of rotordynamic phenomena that can arise from the fluid-induced effects of a finite number of blades, and, in this section, we shall first examine the more general class of phenomena.

Consider the typical geometry of an unshrouded impeller of radius R_T and Z_R blades enclosed by a cylindrical housing so that the mean clearance between the blade tips and the housing is δ (figure 10.10). If the impeller is rotating at a frequency, Ω, and whirling at a frequency, ω, with an amplitude, ϵ, then the vector positions of the blade tips at time, t, will be given by

$$x + jy = z = R_T e^{j(\Omega t + 2\pi n/Z_R)} + \epsilon e^{j\omega t} \quad \text{for } n = 1 \text{ to } Z_R \tag{10.68}$$

where the center of the housing is the origin of the (x, y) coordinate system. It follows that the clearance at each blade tip is $R_T + \delta - |z|$ which, to first order in ϵ, is δ^* where

$$\delta^* = \delta - \epsilon \cos\theta_n, \quad n = 1 \text{ to } Z_R \tag{10.69}$$

and where, for convenience, $\theta_n = \Omega t - \omega t + 2\pi n/Z_R$.

Next, the most general form of the force, F^*, acting on the tip of the blade is

$$F^* = F e^{j(\frac{\pi}{2}+\alpha)} e^{j(\Omega t + 2\pi n/Z_R)} \tag{10.70}$$

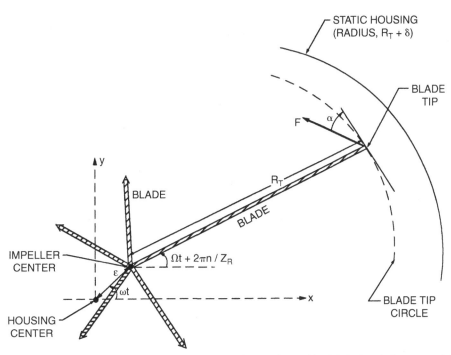

Figure 10.10. Schematic of the position of an axial flow turbomachine blade tip relative to a static housing as a result of the combination of rotational and whirl motion (details shown for only one of the Z_R blade tips).

where the functional forms of the force magnitude, F, and its inclination relative to the blade, α (see figure 10.10), can, for the moment, remain unspecified. The total rotordynamic forces, F_n^* and F_t^*, acting on the impeller are then obtained by appropriate summation of the individual tip forces, F^*, followed by conversion to the rotating frame. Nondimensionalizing the result, one then finds

$$F_n + j F_t = \frac{\left[\displaystyle\sum_{n=1}^{Z_R} j e^{j\alpha} e^{j\theta_n} F\right]_{\text{AVERAGE}}}{\pi \rho \Omega^2 R_T L \epsilon} \tag{10.71}$$

where the quantity in square brackets is averaged over a large time. This general result may then be used, with various postulated relations for F and α, to investigate the resulting rotordynamic effects.

One choice of the form of F and α corresponds to the Alford effect. Alford (1965) surmised that the fluid force acting normal to each blade ($\alpha = 0$ or π) would vary according to the instantaneous tip clearance of that blade. Specifically, he argued that an increase in the clearance, $\epsilon \cos\theta_n$, would produce a proportionate decrease in the normal force, or

$$F = F_0 + \mathcal{K}\epsilon \cos\theta_n \tag{10.72}$$

where F_0 is the mean, time-averaged force normal to each blade and $\mathcal{K}$ is the factor of proportionality. Moreover, for a pump $\alpha = \pi$, and for a turbine $\alpha = 0$. Substituting these values into equation 10.71, one obtains

$$F_n = 0; \quad F_t = \mp \mathcal{K} Z_R / 2\pi \rho \Omega^2 R_T L \qquad (10.73)$$

where the upper sign refers to the pump case and the lower to the turbine case. It follows that the Alford effect in pumps is stabilizing for positive whirl, and destabilizing for negative whirl. In a turbine the reverse is true, and the destabilizing forces for positive whirl can be quite important in the rotordynamics of some turbines.

As a second, but more theoretical example, consider the added mass effect that occurs when a blade tip approaches the casing and squeezes fluid out from the intervening gap. Such a flow would manifest a force on the blade proportional to the acceleration $d^2\delta^*/dt^2$, so that

$$\alpha = \pi/2; \quad F = -\mathcal{K}\frac{d^2\delta^*}{dt^2} \qquad (10.74)$$

where $\mathcal{K}$ is some different proportionality factor. It follows from equation 10.71 that, in this case,

$$F_n = \frac{\mathcal{K}}{2\pi\rho R_T L}\left(1 - \frac{\omega}{\Omega}\right)^2; \quad F_t = 0 \qquad (10.75)$$

This positive normal force is a Bernoulli effect, and has the same basic form as the Bernoulli effect for the whirl of a plane cylinder (see section 10.3).

Other tip clearance flow effects, such as those due to viscous or frictional effects, can be investigated using the general result in equation 10.71, as well as appropriate choices for α and F.

10.10 Steady Radial Forces

We now change the focus of attention back to pumps, and, more specifically, to the kinds of radial and rotordynamic forces which may be caused by the flow through and around an impeller. Unlike some of the devices discussed in the preceding sections, the flow through a pump can frequently be nonaxisymmetric and so can produce a mean radial force that can be of considerable importance. The bearings must withstand this force, and this can lead to premature bearing wear and even failure. Bearing deflection can also cause displacement of the axis of rotation of the impeller, that may, in turn, have deleterious effects upon hydraulic performance. The existence of radial forces, and attempts to evaluate them, date back to the 1930s (see Stepanoff's comment in Biheller 1965) or earlier.

The nonaxisymmetries and, therefore, the radial forces depend upon the geometry of the diffuser and/or volute as well as the flow coefficient. Measurements of radial forces have been made with a number of different impeller/diffuser/volute combinations by Agostonelli *et al.* (1960), Iverson *et al.* (1960), Biheller (1965), Grabow (1964), and Chamieh *et al.* (1985), among others. Stepanoff (1957) proposed an empirical formula for the magnitude of the nondimensional radial force,

$$|F_0| = (F_{0x}^2 + F_{0y}^2)^{\frac{1}{2}} = 0.229\psi \left\{ 1 - (Q/Q_D)^2 \right\} \qquad (10.76)$$

for centrifugal pumps with spiral volutes, and

$$|F_0| = 0.229\psi \, Q/Q_D \qquad (10.77)$$

for collectors with uniform cross-sectional area. Both formulae yield radial forces that have the correct order of magnitude; however, measurements show that the forces also depend on other geometric features of the impeller and its casing.

Some typical nondimensional radial forces obtained experimentally by Chamieh *et al.* (1985) for the Impeller X/Volute A combination (see section 2.8) are shown in figure 10.11 for a range of speeds and flow coefficients. First note that, as anticipated in the nondimensionalization, the radial forces do indeed scale with the square of the impeller speed. This implies that, at least within the range of rotational speeds used for these experiments, the Reynolds number effects on the radial forces are minimal. Second, focusing on Chamieh's data, it should be noted that the "design" objective that Volute A be well matched to Impeller X appears to be satisfied at a flow coefficient, ϕ_2, of 0.092 where the magnitude of the radial force appears to vanish.

Other radial force data are presented in figure 10.12. The centrifugal pump tested by Agostinelli, Nobles and Mockeridge (1960) had a specific speed, N_D, of 0.61, and was similar to that of Chamieh *et al.* (1985). On the other hand, the pump tested by Iversen, Rolling and Carlson (1960) had a much lower specific speed of 0.36, and the data of figure 10.12 indicates that their impeller/volute combination is best matched at a flow coefficient of about 0.06. The data of Domm and Hergt (1970) is for a volute similar to Volute A and, while qualitatively similar to the other data, has a significantly smaller magnitude than the other three sets of data. The reasons for this are not clear.

The dependence of the radial forces on volute geometry is illustrated in figure 10.13 from Chamieh *et al.* (1985) which presents a comparison of the magnitude of the force on Impeller X due to Volute A with the magnitude of the force due to a circular volute with a circumferentially uniform cross-sectional area. In theory, this second volute could only be well-matched at zero flow rate; note that the results do exhibit a minimum at shut-off. Figure 10.13 also illustrates one of the compromises that a designer may have to make. If the objective were to minimize the radial force at a single flow rate, then a well-designed spiral volute would be appropriate. On the other hand, if the objective were to minimize the force over a wide range of flow rates, then a quite different design, perhaps even a constant area volute, might be more effective. Of course, a comparison of the hydraulic performance would also have to be

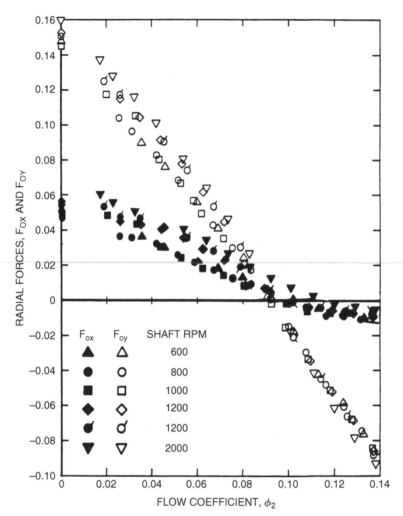

Figure 10.11. Radial forces for the centrifugal Impeller X/Volute A combination as a function of shaft speed and flow coefficient (Chamieh *et al.* 1985).

made in evaluating such design decisions. Note from figure 7.1 that the spiral volute is hydraulically superior up to a flow coefficient of 0.10 above which the results are circular volute is superior.

As further information on the variation of the magnitude of the radial forces in different types of pump, we include figure 10.14, taken from KSB (1975), which shows how F_0/ψ may vary with specific speed and flow rate for a class of volute pumps. The magnitudes of the forces shown in this figure are larger than those of figure 10.12. We should also note that the results of Jery and Franz (1982) indicate that the presence of diffuser vanes (of typical low solidity) between the impeller discharge and the volute has relatively little effect on the radial forces.

It is also important to recognize that small changes in the location of the impeller within the volute can cause large changes in the radial forces. This gradient of forces

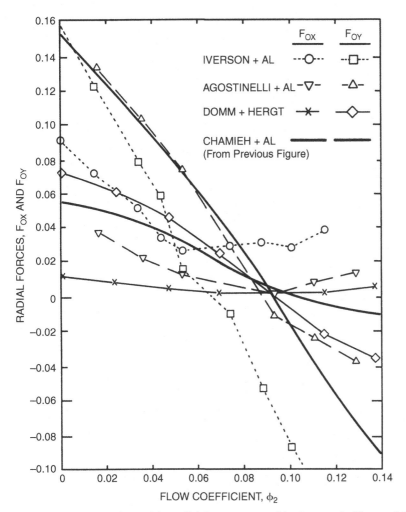

Figure 10.12. Comparison of the radial forces measured by Iverson, Rolling, and Carlson (1960) on a pump with a specific speed, N_D, of 0.36, by Agostinelli, Nobles, and Mockeridge (1960), on a pump with $N_D = 0.61$, by Domm and Hergt (1970), and by Chamieh *et al.* (1985) on a pump with $N_D = 0.57$.

is represented by the hydrodynamic stiffness matrix, $[K]$ (see section 10.2), for which data will be presented in the context of the rotordynamic coefficients. The dependence of the radial force on the impeller position also implies that, for a given impeller/volute combination at a particular flow coefficient, there exists a particular location of the axis of impeller rotation for which the radial force is zero. As an example, the locus of zero radial force positions for the Impeller X/Volute A combination is presented in figure 10.15. Note that this location traverses a distance of about 10% of the impeller radius as the flow rate increases from zero to a flow coefficient of 0.14.

Visualizing the centrifugal pump impeller as a control volume, one can recognize three possible contributions to the radial force. First, circumferential variation in the impeller discharge pressure (or volute pressure) will clearly result in a radial

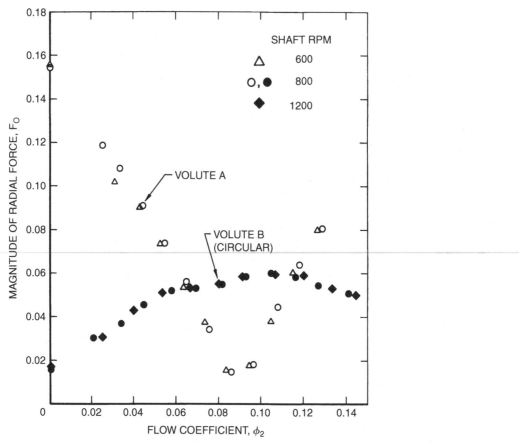

Figure 10.13. Comparison of the magnitude of the radial force (F_0) on Impeller X caused by Volute A and by the circular Volute B with a circumferentially uniform area (Chamieh *et al.* 1985).

force acting on the impeller discharge area. A second contribution could be caused by the leakage flow from the impeller discharge to the inlet between the impeller shroud and the pump casing. Circumferential nonuniformity in the discharge pressure could cause circumferential nonuniformity in the pressure within this shroud-casing gap, and therefore a radial force acting on the exterior of the pump shroud. For convenience, we shall term this second contribution the leakage flow contribution. Third, a circumferential nonuniformity in the flow rate out of the impeller would imply a force due to the nonuniformity in the momentum flux out of the impeller. This potential third contribution has not been significant in any of the studies to date. Both the first two contributions appear to be important.

In order to investigate the origins of the radial forces, Adkins and Brennen (1988) (see also Brennen *et al.* 1986) made measurements of the pressure distributions in the volute, and integrated these pressures to evaluate the contribution of the discharge pressure to the radial force. Typical pressure distributions for the Impeller X/Volute A combination (with the flow separation rings of figure 10.17 installed) are presented

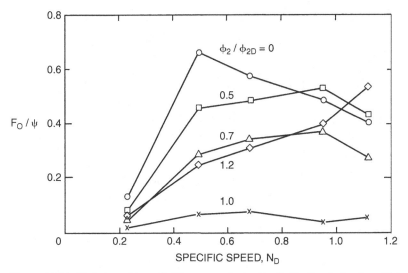

Figure 10.14. Variation of the radial force magnitude, F_0, divided by the head coefficient, ψ, as a function of specific speed, N_D, and flow for a class of volute casing pumps (adapted from KSB 1975).

in figure 10.16 for three different flow coefficients. Minor differences occur in the pressures measured in the front sidewall of the volute at the impeller discharge (front taps) and those in the opposite wall (back taps).

The experimental measurements in figure 10.18 are compared with theoretical predictions based on an analysis that matches a guided impeller flow model with a one-dimensional treatment of the flow in the volute. This same theory was used to calculate rotordynamic matrices and coefficients presented in section 10.12. In the present context, integration of the experimental pressure distributions yielded radial forces in good agreement with both the overall radial forces measured using the force balance and the theoretical predictions of the theory. These results demonstrate that it is primarily the circumferential nonuniformity in the pressure at the impeller discharge that generates the radial force. The theory clearly demonstrates that the momentum flux contribution is negligible.

The leakage flow from the impeller discharge, between the impeller shroud and the pump casing, and back to the pump inlet does make a significant contribution to the radial force. Figure 10.17 is a schematic of the impeller, volute, and casing used in the experiments of Chamieh *et al.* (1985) and Adkins and Brennen (1988), as well as for the rotordynamic measurements discussed later. Adkins and Brennen obtained data with and without the obstruction at the entrance to the leakage flow labeled "flow separation rings". The data of figures 10.16 and 10.18 were taken with these rings installed (whereas Chamieh's data was taken without the rings). The measurements showed that, in the absence of the rings, the nonuniformity in the impeller discharge pressure caused significant nonuniformity in the pressure in the leakage annulus, and, therefore, a significant contribution from the leakage flow to the radial force. This

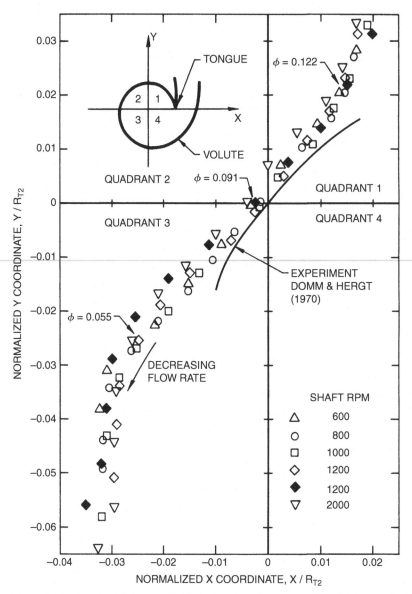

Figure 10.15. Locus of the zero radial force locations for the Impeller X/Volute A combination (Chamieh *et al.* 1985) compared with that from the data of Domm and Hergt (1970).

was not the case once the rings were installed, for the rings effectively isolated the leakage annulus from the impeller discharge nonuniformity. However, a compensating mechanism exists which causes the total radial force in the two cases to be more or less the same. The increased leakage flow without the rings tends to relieve some of the pressure nonuniformity in the impeller discharge, thus reducing the contribution from the impeller discharge pressure distribution.

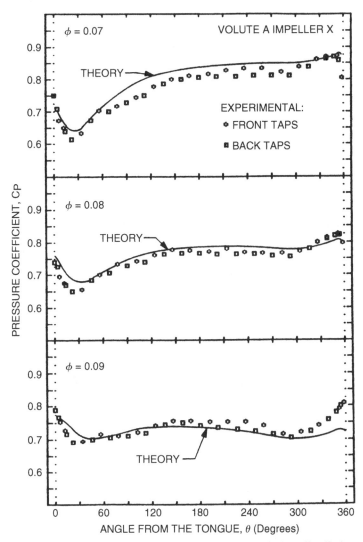

Figure 10.16. Circumferential pressure distributions in the impeller discharge for the Impeller X/Volute A combination at three different flow rates. Also shown are the theoretical pressure distributions of Adkins and Brennen (1988).

A number of other theoretical models exist in the literature. The analysis of Lorett and Gopalakrishnan (1983) is somewhat similar in spirit to that of Adkins and Brennen (1988). Earlier analyses, such as those of Domm and Hergt (1970) and Colding-Jorgensen (1979), were based on modeling the impeller by a source/vortex within the volute and solutions of the resulting potential flow. They represent too much of a departure from real flows to be of much applicability.

Finally, we note that the principal focus of this section has been on radial forces caused by circumferential nonuniformity in the discharge conditions. It must be clear that nonuniformities in the inlet flow due, for example, to bends in the suction piping

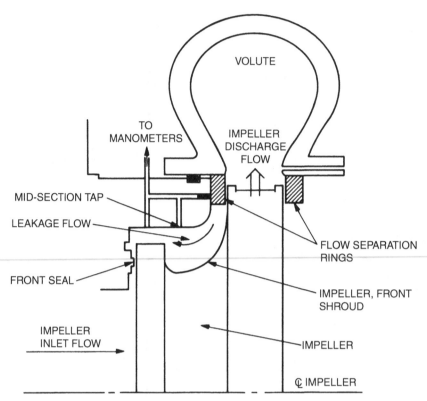

Figure 10.17. Schematic of the Impeller X/Volute A arrangement used for the experiments of Chamieh *et al.* (1985) and Adkins and Brennen (1988).

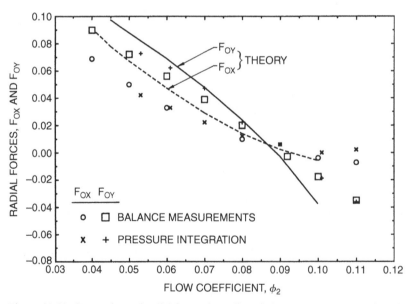

Figure 10.18. Comparison of radial forces from direct balance measurements, from integration of measured pressures, and from theory for the Impeller X/Volute A combination (from Adkins and Brennen 1988).

are also likely to generate radial forces. As yet, such forces have not been investigated. Moreover, it seems reasonable to suggest that inlet distortion forces are more likely to be important in axial inducers or pumps than in centrifugal pumps.

10.11 Effect of Cavitation

Franz *et al.* (1990) (see also Brennen *et al.* 1988) have made measurements of the radial forces for the Impeller X/Volute A combination under cavitating conditions. These studies show that any loss of head can also cause major changes in the magnitude and direction of the radial force. This is illustrated in figure 10.19, where the cavitation performance is juxtaposed with the variation in the radial forces for three different flow coefficients. Note that the radial force changes when the head rise across the pump is affected by cavitation. Note also that the changes in the radial forces are large, in some instances switching direction by 180° while the flow rate remains the same. This result may be of considerable significance since pumps operating near breakdown often exhibit fluctuations in which the operating point moves back and forth over the knee of the cavitation performance curve. According to figure 10.19, such performance fluctuations would result in large fluctuating forces that could well account for the heavy vibration and rough running that is usually manifest by a pump operating under cavitating conditions.

10.12 Centrifugal Pumps

Rotordynamic forces in a centrifugal pump were first measured by Hergt and Krieger (1969–70), Ohashi and Shoji (1984b) and Jery *et al.* (1985). Typical data for the dimensionless normal and tangential forces, F_n and F_t, as a function of the frequency ratio, ω/Ω, are presented in figure 10.20 for the Impeller X/Volute A combination. The curve for Impeller X is typical of a wide range of results at different speeds, flow coefficients, and with different impellers and volutes. Perhaps the most significant feature of these results is that there exists a range of whirl frequencies for which the tangential force is whirl destabilizing. A positive F_t at negative whirl frequencies opposes the whirl motion, and is, therefore, stabilizing, and fairly strongly so since the forces can be quite large in magnitude. Similarly, at large, positive frequency ratios, the F_t is negative and is also stabilizing. However, between these two stabilizing regions, one usually finds a regime at small positive frequency ratios where F_t is positive and therefore destabilizing.

As is illustrated by figure 10.20, the variation of F_n and F_t with the whirl frequency ratio, ω/Ω, can be represented quite accurately by the quadratic expressions of equations 10.13 and 10.14 (this is not true for axial flow pumps, as will be discussed later). The rotordynamic coefficients, obtained from data like that of figure 10.20 for a wide variety of speeds, flow rates, and impeller, diffuser, and volute geometries,

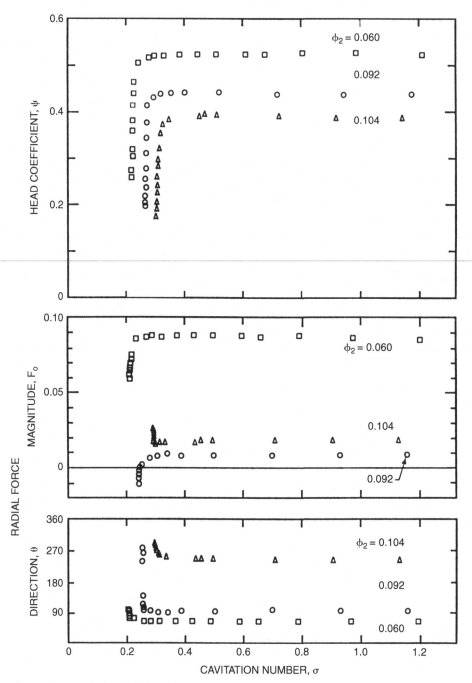

Figure 10.19. Variation of the head and radial force (magnitude, F_0, and direction, θ, measured from the cutwater) with cavitation number, σ, for Impeller X/Volute A at three flow coefficients and at 3000 rpm (from Franz *et al.* 1990).

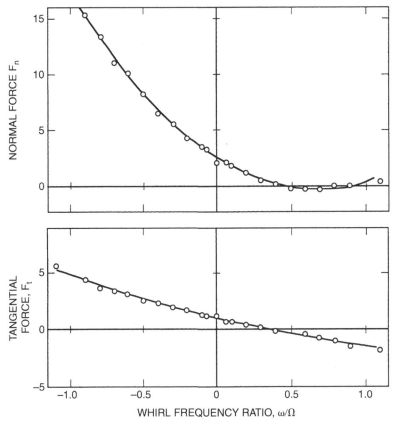

Figure 10.20. Typical rotordynamic forces, F_n and F_t, as a function of whirl frequency ratio, ω/Ω, for the Impeller X/Volute A combination running at $1000\ rpm$ and a flow coefficient of $\phi_2 = 0.092$ (from Jery *et al.* 1985).

are given in table 10.2 (adapted from Jery *et al.* 1985). Note, first, some of the general characteristics of these coefficients. The direct stiffness, K, is always negative because of the Bernoulli effect (see section 10.3). The cross-coupled stiffness, k, is always positive, and is directly connected to the positive values of F_t at low positive whirl frequency ratios; consequently, k is a measure of the destabilizing effect of the fluid. The direct damping, C, is positive, but usually less than half of the value of the cross-coupled damping, c. Note that the value of k/C is usually a fairly accurate measure of the whirl frequency ratio corresponding to the upper bound of the destabilizing interval of whirl frequency ratios. From table 10.2 the values of k/C, for actual impellers with volutes and with nonzero flow, range from 0.25 to 0.40, so the range of subsynchronous speeds, for which these fluid forces are destabilizing, can be quite large. Resuming the summary of the rotordynamic coefficients, note that the cross-coupled added mass, m, is small in comparison with the direct added mass, M, and can probably be neglected in many applications. Note that, since the direct added mass is converted to dimensional form by $\pi \rho R_{T2}^2 B_2$, it follows that typical values of

Table 10.2. *Rotordynamic coefficients for various centrifugal pump configurations (from Jery et al. 1985). Volute E is a 17-bladed diffuser with a spiral volute. Volutes D, F, G, and H are spiral volutes fitted with zero, 6, 6, and 12 vanes respectively. Impeller Y is a 6-bladed impeller. Impeller S is a solid mass with the same external profile as Impeller X*

Impeller/Volute	rpm	ϕ_2	K	k	C	c	M	m
Imp.X/Volute A	500	0.092	−2.51	1.10	3.14	7.91	6.52	−0.52
	1000	0.092	−2.61	1.12	3.28	8.52	6.24	−0.53
	1500	0.092	−2.47	0.99	3.00	8.71	6.87	−0.87
	2000	0.092	−2.64	1.15	2.91	9.06	7.02	−0.67
Imp.X/Volute E	1000	0	−1.64	0.14	3.40	7.56	6.83	0.68
	1000	0.060	−2.76	1.02	3.74	9.53	6.92	−1.01
	1000	0.092	−2.65	1.04	3.80	8.96	6.60	−0.90
	1000	0.145	−2.44	1.16	4.11	7.93	6.20	−0.55
Imp.X/none	1000	0.060	−0.55	0.67	1.24	3.60	4.38	1.68
Imp.X/Volute D	1000	0.060	−2.86	1.12	2.81	9.34	6.43	−0.15
Imp.X/Volute F	1000	0.060	−3.40	1.36	3.64	9.51	6.24	−0.72
Imp.X/Volute G	1000	0.060	−3.34	1.30	3.42	9.11	5.75	−0.39
Imp.X/Volute H	1000	0.060	−3.42	1.33	3.75	10.34	7.24	−0.65
Imp.Y/Volute E	1000	0.092	−2.81	0.85	3.34	8.53	5.50	−0.74
Imp.S/Volute A	1000		−0.42	0.41	1.87	3.81	6.54	−0.04

the added mass, M, are equivalent to the mass of about six such cylinders, or about five times the volume of liquid inside the impeller.

Now examine the variations in the values of the rotordynamic coefficients in table 10.2. The first series of data clearly demonstrates that the nondimensionalization has satisfactorily accounted for the variation with rotational speed. Any separate effect of Reynolds number does not appear to occur within the range of speeds in these experiments. The second series in table 10.2 illustrates the typical variations with flow coefficient. Note that, apart from the stiffness at zero flow, the coefficients are fairly independent of the flow coefficient. The third series utilized diffusers with various numbers and geometries of vanes inside the same volute. The presence of vanes appears to cause a slight increase in the stiffness; however, the number and type of vanes do not seem to matter. Note that, in the absence of any volute or diffuser, all of the coefficients (except m) are substantially smaller. Ohashi and Shoji (1984b) made rotordynamic measurements within a much larger volute than any in table 10.2; consequently their results are comparable with those given in table 10.2 for no volute. On the other hand, Bolleter, Wyss, Welte, and Sturchler (1985, 1987) report rotordynamic coefficients very similar in magnitude to those of table 10.2.

The origins of the rotordynamic forces in typical centrifugal pumps have been explored by Jery *et al.* (1985) and Adkins and Brennen (1988), among others. In order to explore the effect of the discharge-to-suction leakage flow between the shroud and the casing, Jery *et al.* (1985) compared the rotordynamic forces generated by the Impeller X/Volute A combination with those generated in the same housing by a dummy impeller (Impeller S) with the same exterior profile as Impeller X. A pressure difference was externally applied in order to simulate the same inlet to discharge static pressure rise, and, therefore, produce a leakage flow similar to that in the Impeller X experiments. As in the case of the radial forces, we surmise that unsteady circumferential pressure differences on the impeller discharge and in the leakage flow can both contribute to the rotordynamic forces on an impeller. As can be seen from the coefficients listed in table 10.2, the rotordynamic forces with the dummy impeller represented a substantial fraction of those with the actual impeller. We conclude that the contributions to the rotordynamic forces from the unsteady pressures acting on the impeller discharge and those from the unsteady pressures in the leakage flow acting on the shroud are both important and must be separately investigated and evaluated.

We focus first on the impeller discharge contribution. Adkins and Brennen (1988) used an extension of the theoretical model described briefly in section 10.10 to evaluate the rotordynamic forces acting on the impeller discharge. They also made measurements of the forces for an Impeller X/Volute A configuration in which the pump casing structure external to the shroud was removed in order to minimize any contributions from the leakage flow. The resulting experimental and theoretical values of F_n and F_t are presented in figure 10.21. First note that these values are significantly smaller than

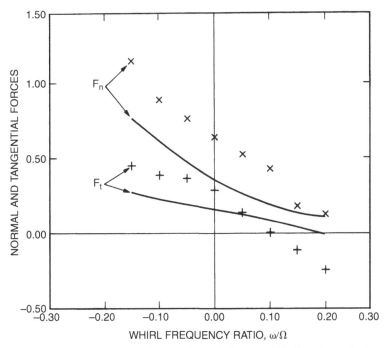

Figure 10.21. Comparison of the rotordynamic force contributions due to the impeller discharge pressure variations as predicted by the theory of Adkins and Brennen (1988) (solid lines) with experimental measurements using Impeller X and Volute A (at $\phi_2 = 0.092$) but with the casing surrounding the front shroud removed to minimize the leakage flow contributions.

those of figure 10.20, indicating that the impeller discharge contributions are actually smaller than those from the leakage flow. Second note that the theory of Adkins and Brennen (1988) provides a reasonable estimate of the impeller discharge contribution to the rotordynamic forces, at least within the range of whirl frequencies examined.

Using the Impeller X/Volute A configuration, Adkins and Brennen also made experimental measurements of the pressure distributions in the impeller discharge flow and in the leakage flow. These measurements allowed calculations of the stiffnesses, $K = F_n(0)$ and $k = F_t(0)$. The results indicated that the leakage flow contributes about 70% of K and about 40% of k; these fractional contributions are similar to those expected from a comparison of figures 10.20 and 10.21.

About the same time, Childs (1987) used the bulk-flow model described in section 10.7 to evaluate the contributions to the rotordynamic forces from the discharge-to-suction leakage flow. While his results exhibit some peculiar resonances not yet observed experimentally, the general magnitude and form of Childs results are consistent with the current conclusions. More recently, Guinzberg et al. (1990) have made experimental measurements for a simple leakage flow geometry that clearly confirm the importance of the rotordynamic effects caused by these flows. They also demonstrate the variations in the leakage flow contributions with the geometry of the leakage path, the leakage flow rate and the swirl in the flow at the entrance to the leakage path.

It is important to mention previous theoretical investigations of the rotordynamic forces acting on impellers. A number of the early models (Thompson 1978, Colding-Jorgensen 1979, Chamieh and Acosta 1981) considered only quasistatic perturbations from the mean flow, so that only the stiffness can be evaluated. Ohashi and Shoji (1984a) (see also Shoji and Ohashi 1980) considered two-dimensional, inviscid and unseparated flow in the impeller, and solved the unsteady flow problem by singularity methods. Near the design flow rate, their results compare well with their experimental data, but at lower flows the results diverge. More recently, Tsujimoto et al. (1988) have included the effects of a volute; their two-dimensional analysis yielded good agreement with the measurements by Jery et al. (1985) on a two-dimensional impeller.

Finally, in view of the significant effect of cavitation on the radial forces (section 10.10), it is rather surprising to find that the effect of cavitation on the rotordynamic forces in centrifugal pumps seems to be quite insignificant (Franz et al. 1990).

10.13 Moments and Lines of Action

Some data on the steady bending moments, M_{ox} and M_{oy}, and on the rotordynamic moments

$$M_n = B_{xx} = B_{yy}; \quad M_t = B_{yx} = -B_{xy} \tag{10.78}$$

have been presented by Franz et al. (1990) and Miskovish and Brennen (1992). This data allows evaluation of the axial location of the lines of action of the corresponding

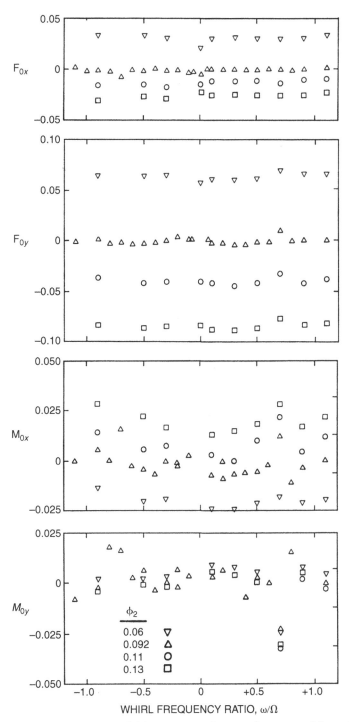

Figure 10.22. Steady radial forces, F_{ox} and F_{oy}, and moments, M_{ox} and M_{oy}, for Impeller X/Volute A at a speed of 1000 rpm and various flow coefficients as indicated (from Miskovish and Brennen 1992).

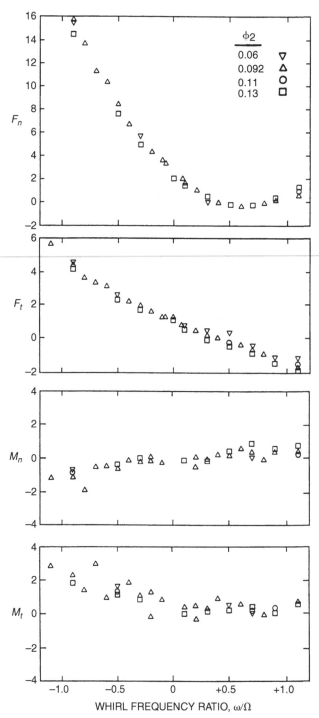

Figure 10.23. Normal and tangential rotordynamic forces, F_n and F_t, and moments, M_n and M_t, for Impeller X/Volute A at 1000 rpm and various flow coefficients as indicated (from Miskovish and Brennen 1992).

radial and rotordynamic forces. Apart from its intrinsic value, knowledge of the line of action of these forces provides clues as to the origin of the forces.

Typical sets of data taken from Miskovish and Brennen (1992) are presented in figures 10.22 and 10.23. These were obtained for the Impeller X/Volute A combination operating at a speed of 1000 rpm. For convenience, the axial location of the origin of the reference coordinate system has been placed at the center of the impeller discharge. Since the lines of action of the forces are not too far from this location, the moments presented here are small, and, for this reason, the data for the moments is somewhat scattered.

Steady forces and moments are presented for many whirl frequency ratios in figure 10.22. These forces and moments should, of course, be independent of the whirl frequency ratio, and so the deviation of the data points from the mean for a given flow coefficient represents a measure of the scatter in the data. Despite this scatter, the moment data in figure 10.22 does suggest that a nonzero steady moment is present, and that it changes with flow coefficient. The typical location for the line of action of F_o, which this data implies, may be best illustrated by an example. At $\phi = 0.06$, the steady vector force F_o has a magnitude of 0.067 ($F_{ox} \approx 0.03$, $F_{oy} \approx 0.06$) and an angle $\theta_F = 63°$ from the x-axis. The corresponding moment vector has a magnitude of 0.02 and an angle $\theta_M \sim 180°$ from the x-axis. Consequently, the line of action of F_o is an axial distance *upstream* of the origin equal to $0.02 \sin(180 - 63)/0.067 = 0.27$. In other words, the line of action is about a quarter of a discharge radius upstream of the center of the discharge. This is consistent with the previous observation (section 10.10) that the pressures acting on the exterior of the shroud also contribute to the steady radial forces; this contribution displaces the line of action upstream of the center of the discharge.

The data of figure 10.23 could be similarly used to evaluate the lines of action of the rotordynamic forces whose components are F_n and F_t. However, the moments M_n and M_t are small over most of the range of whirl ratios, and lead to lines of action that are less than 0.1 of a radius upstream of the center of the discharge in most cases. This is consistent with other experiments on this same impeller/volute/casing combination that suggest that the shroud force contribution to the rotordynamic matrices is smaller than the impeller discharge contribution in this particular case.

10.14 Axial Flow Inducers

The rotordynamic forces in an unshrouded axial flow pump, or those caused by adding an axial inducer to a centrifugal pump, are less well understood. One of the reasons for this is that the phenomena will depend on the dynamic response of the tip clearance flows, an unsteady flow that has not been studied in any detail. The experimental data that does exist (Franz and Arndt 1986, Arndt and Franz 1986, Karyeaclis *et al.* 1989) clearly show that important and qualitatively different effects are manifest by unshrouded axial flow pumps. These effects were not encountered with shrouded

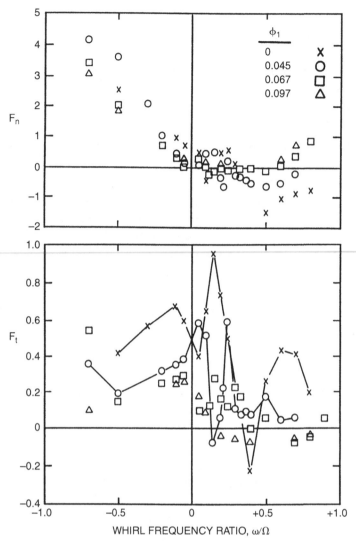

Figure 10.24. Rotordynamic forces for the helical inducer, Impeller VII, for four different flow coefficients (from Arndt and Franz 1986).

centrifugal impellers. They are exemplified by figure 10.24, which presents data on F_n and F_t for the 9° helical inducer, Impeller VII, tested alone at a series of flow coefficients (Arndt and Franz 1986). At the higher flow coefficients, the variation of F_n and F_t with whirl frequency ratio, ω/Ω, is similar to the centrifugal pump data. However, as the flow coefficient is decreased, somewhat pathological behavior begins to appear in the values of F_t (and to a lesser degree F_n) at small and positive whirl frequency ratios. This culminates in extremely complicated behavior at shut-off (zero flow) in which F_t changes sign several times for positive whirl frequency ratios, implying several separate regions of destabilizing fluid-induced rotordynamic effect. Note that the maximum values of F_t that were recorded, are large, and could well be responsible for significant vibration in an axial flow pump or inducer. Similar pathological rotordynamic effects were encountered with all the axial inducers tested, including the inducer/impeller combination represented by the high pressure LOX pump in the Space Shuttle Main Engine (Franz and Arndt 1986). However, the details in the variations of F_t with ω/Ω differed from one inducer to another.

Finally, we should note that the current codes for rotordynamic investigations are not well adapted to deal with deviations from the quadratic forms for F_n^* and F_t^* given in equations 10.13 and 10.14. Consequently, more remains to be done in terms of rotordynamic analysis before the implications of such complex frequency-dependent behavior of F_n^* and F_t^* become clear.

Bibliography

ASTM (Amer. Soc. for Testing and Materials). (1967). Erosion by cavitation or impingement. *ASTM STP408*.

Abbott, M.B. (1966). *An introduction to the method of characteristics.* American Elsevier, New York.

Abramson, H.N. (1969). Hydroelasticity: a review of hydrofoil flutter. *Appl. Mech. Rev.*, **22**, No. 2, 115–121.

Acosta, A.J. (1955). A note on partial cavitation of flat plate hydrofoils. *Calif. Inst. of Tech. Hydro. Lab. Rep. E-19.9.*

Acosta, A.J. and Bowerman, R.D. (1957). An experimental study of centrifugal pump impellers. *Trans. ASME*, **79**, 1821–1839.

Acosta, A.J. (1958). An experimental study of cavitating inducers. *Proc. Second ONR Symp. on Naval Hydrodyn.*, **ONR/ACR-38**, 533–557.

Acosta, A.J. and Hollander, A. (1959). Remarks on cavitation in turbomachines. *Calif. Inst. of Tech. Rep. E-79.3.*

Acosta, A.J. (1960). Cavitating flow past a cascade of circular arc hydrofoils. *Calif. Inst. of Tech. Hydro. Lab. Rep. E-79.2.*

Acosta, A.J. and DeLong, R.K. (1971). Experimental investigation of non-steady forces on hydrofoils oscillating in heave. *Proc. IUTAM Symp. on non-steady flow of water at high speeds, Leningrad, USSR*, 95–104.

Acosta, A.J. (1973). Hydrofoils and hydrofoil craft. *Ann. Rev. Fluid Mech.*, **5**, 161–184.

Acosta, A.J. and Parkin, B.R. (1975). Cavitation inception—a selective review. *J. Ship Res.*, **19**, No. 4, 193–205.

Adamczyk, J.J. (1975). The passage of a distorted velocity field through a cascade of airfoils. *Proc. Conf. on Unsteady Phenomena in Turbomachinery, AGARD Conf. Proc. No. 177.*

Adkins, D.R. and Brennen, C.E. (1988). Analyses of hydrodynamic radial forces on centrifugal pump impellers. *ASME J. Fluids Eng.*, **110**, No. 1, 20–28.

Agostinelli, A., Nobles, D., and Mockridge, C.R. (1960). An experimental investigation of radial thrust in centrifugal pumps. *ASME J. Eng. for Power*, **82**, 120–126.

Alford, J.S. (1965). Protecting turbomachinery from self-excited rotor whirl. *ASME. J. Eng. for Power*, **87**, 333–344.

Amies, G., Levek, R., and Struesseld, D. (1977). Aircraft hydraulic systems dynamic analysis. Volume II. Transient analysis (HYTRAN). *Wright-Patterson Air Force Base Technical Report AFAPL-TR-76-43*, **II**.

Amies, G. and Greene, B. (1977). Aircraft hydraulic systems dynamic analysis. Volume IV. Frequency response (HSFR). *Wright-Patterson Air Force Base Technical Report AFAPL-TR-76-43*, **IV**.

Anderson, D.A., Blade, R.J., and Stevens, W. (1971). Response of a radial-bladed centrifugal pump to sinusoidal disturbances for non-cavitating flow. *NASA TN D-6556.*

Anderson, H.H. (Undated). *Centrifugal pumps*. The Trade and Technical Press Ltd., Crown House, Morden, England.

Anderson, H.H. (1955). Modern developments in the use of large single-entry centrifugal pumps. *Proc. Inst. Mech. Eng.*, **169**, 141–161.

Arakeri, V.H. and Acosta, A.J. (1974). Viscous effects in the inception of cavitation on axisymmetric bodies. *ASME J. Fluids Eng.*, **95**, No. 4, 519–528.

Arakeri, V.H. (1979). Cavitation inception. *Proc. Indian Acad. Sci.*, **C2**, Part 2, 149–177.

Arakeri, V.H. and Shangumanathan, V. (1985). On the evidence for the effect of bubble interference on cavitation noise. *J. Fluid Mech.*, **159**, 131–150.

Arndt, N. and Franz, R. (1986). Observations of hydrodynamic forces on several inducers including the SSME LPOTP. *Calif. Inst. of Tech., Div. Eng. and Appl. Sci., Report No. E249.3*.

Arndt, N., Acosta, A.J., Brennen, C.E., and Caughey, T.K. (1989). Rotor-stator interaction in a diffuser pump. *ASME J. of Turbomachinery*, **111**, 213–221.

Arndt, N., Acosta, A.J., Brennen, C.E., and Caughey, T.K. (1990). Experimental investigation of rotor-stator interaction in a centrifugal pump with several vaned diffusers. *ASME J. of Turbomachinery*, **112**, 98–108.

Arndt, R.E.A. and Ippen, A.T. (1968). Rough surface effects on cavitation inception. *ASME J. Basic Eng.*, **90**, 249–261.

Arndt, R.E.A. and Keller, A.P. (1976). Free gas content effects on cavitation inception and noise in a free shear flow. *Proc. IAHR Symp. on two-phase flow and cavitation in power generation*, 3–16.

Arndt, R.E.A. (1981). Cavitation in fluid machinery and hydraulic structures. *Ann. Rev. Fluid Mech.*, **13**, 273–328.

Badowski, H.R. (1969). An explanation for instability in cavitating inducers. *Proc. 1969 ASME Cavitation Forum*, 38–40.

Badowski, H.R. (1970). Inducers for centrifugal pumps. *Worthington Canada, Ltd., Internal Report*.

Balje, O.E. (1981). *Turbomachines. A guide to design, selection and theory*. John Wiley and Sons, New York.

Batchelor, G.K. (1967). *An introduction to fluid dynamics*. Cambridge Univ. Press.

Benjamin, T.B. and Ellis, A.T. (1966). The collapse of cavitation bubbles and the pressures thereby produced against solid boundaries. *Phil. Trans. Roy. Soc., London, Ser. A*, **260**, 221–240.

Betz, A. and Petersohn, E. (1931). Application of the theory of free jets. *NACA TM No. 667*.

Bhattacharyya, A., Acosta, A.J., Brennen, C.E., and Caughey, T.K. (1993). Observations on off-design flows in non-cavitating inducers. *Proc. ASME Symp. on Pumping Machinery - 1993*, **FED-154**, 135–141.

Biesheuvel, A. and van Wijngaarden, L. (1984). Two-phase flow equations for a dilute dispersion of gas bubbles in liquid. *J. Fluid Mech.*, **148**, 301–318.

Biheller, H.J. (1965). Radial force on the impeller of centrifugal pumps with volute, semi-volute and fully concentric casings. *ASME J. Eng. for Power, July 1965*, 319–323.

Billet, M.L. (1985). Cavitation nuclei measurements—a review. *Proc. 1985 Cavitation and Multiphase Flow Forum*, **FED-23**, 31–38.

Binder, R.C. and Knapp, R.T. (1936). Experimental determinations of the flow characteristics in the volutes of centrifugal pumps. *Trans. ASME*, **58**, 649–661.

Birkhoff, G. and Zarantonello, E.M. (1957). *Jets, wakes and cavities*. Academic Press, NY.

Black, H.F. (1969). Effects of hydraulic forces in annular pressure seals on the vibrations of centrifugal pump rotors. *J. Mech. Eng. Sci.*, **11**, No. 2, 206–213.

Black, H.F. and Jensen, D.N. (1970). Dynamic hybrid properties of annular pressure seals. *Proc. J. Mech. Eng.*, **184**, 92–100.

Blake, F.G. (1949). The onset of cavitation in liquids: I. *Acoustics Res. Lab., Harvard Univ., Tech. Memo. No. 12*.

Blake, W.K., Wolpert, M.J., and Geib, F.E. (1977). Cavitation noise and inception as influenced by boundary-layer development on a hydrofoil. *J. Fluid Mech.*, **80**, 617–640.

Blake, J.R. and Gibson, D.C. (1987). Cavitation bubbles near boundaries. *Ann. Rev. Fluid Mech.*, **19**, 99–124.

Bolleter, U., Wyss, A., Welte, I., and Sturchler, R. (1985). Measurements of hydrodynamic interaction matrices of boiler feed pump impellers. *ASME Paper No. 85-DET-148.*

Bolleter, U., Wyss, A., Welte, I., and Sturchler, R. (1987). Measurement of hydrodynamic matrices of boiler feed pump impellers. *ASME J. Vibrations, Stress and Reliability in Design*, **109**, 144–151.

Braisted, D.M. (1979). Cavitation induced instabilities associated with turbomachines. *Ph.D. Thesis, California Institute of Technology, Pasadena, Ca.*

Braisted, D.M. and Brennen, C.E. (1980). Auto-oscillation of cavitating inducers. In *Polyphase Flow and Transport Technology*, (ed: R.A. Bajura), ASME Publ., New York, 157–166.

Brennen, C.E. (1973). The dynamic behavior and compliance of a stream of cavitating bubbles. *ASME J. Fluids Eng.*, **95**, 533–542.

Brennen, C. (1976). On the flow in an annulus surrounding a whirling cylinder. *J. Fluid Mech.*, **75**, 173–191.

Brennen, C.E. (1978). Bubbly flow model for the dynamic characteristics of cavitating pumps. *J. Fluid Mech.*, **89**, Part 2, 223–240.

Brennen, C.E. (1994). *Cavitation and bubble dynamics.* Oxford Univ. Press.

Brennen, C.E. and Acosta, A.J. (1973). Theoretical, quasistatic analyses of cavitation compliance in turbopumps. *J. of Spacecraft and Rockets*, **10**, No. 3, 175–180.

Brennen, C.E. and Acosta, A.J. (1975). The dynamic performance of cavitating turbopumps. *Proc. Fifth Conference on Fluid Machinery, Akademiai Kiado, Budapest, Hungary*, 121–136.

Brennen, C.E. and Acosta, A.J. (1976). The dynamic transfer function for a cavitating inducer. *ASME J. Fluids Eng.*, **98**, 182–191.

Brennen, C.E., Acosta, A.J., and Caughey, T.K. (1986). Impeller fluid forces. *Proc. NASA Advanced Earth-to-Orbit Propulsion Technology Conference, Huntsville, AL, NASA Conf. Publ. 2436*, 270–295.

Brennen, C.E. and Braisted, D.M. (1980). Stability of hydraulic systems with focus on cavitating pumps. *Proc. 10th Symp. of IAHR, Tokyo*, 255–268.

Brennen, C.E., Franz, R., and Arndt, N. (1988). Effects of cavitation on rotordynamic force matrices. *Proc. NASA Advanced Earth-to-Orbit Propulsion Technology Conference, Huntsville, AL, NASA Conf. Publ. 3012*, 227–239.

Brennen, C.E., Meissner, C., Lo, E.Y., and Hoffman, G.S. (1982). Scale effects in the dynamic transfer functions for cavitating inducers. *ASME J. Fluids Eng.*, **104**, 428–433.

Brennen, C.E., Oey, K., and Babcock, C.D. (1980). On the leading edge flutter of cavitating hydrofoils. *J. Ship Res.*, **24**, No. 3, 135–146.

Breugelmans, F.A.E. and Sen, M. (1982). Prerotation and fluid recirculation in the suction pipe of centrifugal pumps. *Proc. 11th Int. Pump Symp., Texas A&M Univ.*, 165–180.

Brown, F.T. (1967). A unified approach to the analysis of uniform one-dimensional distributed systems. *ASME J. Basic Eng.*, **89**, No. 6, 423–432.

Busemann, A. (1928). Das Förderhöhenverhältnis radialer Kreiselpumpen mit logarithmisch-spiraligen Schaufeln. *Z. angew. Math. u. Mech.*, **8**, 372.

Ceccio, S.L. and Brennen, C.E. (1991). Observations of the dynamics and acoustics of travelling bubble cavitation. *J. Fluid Mech.*, **233**, 633–660. Corrigenda, **240**, 686.

Chahine, G.L. (1982). Cloud cavitation theory. *Proc. 14th ONR Symp. on Naval Hydrodynamics*, 165–194.

Chamieh, D.S. and Acosta, A.J. (1981). Calculation of the stiffness matrix of an impeller eccentrically located within a volute. *Proc. ASME Cavitation and Polyphase Flow Forum*, 51–53.

Chamieh, D. (1983). Forces on a whirling centrifugal pump-impeller. *Ph.D. Thesis, Calif. Inst. of Tech., and Div. of Eng. and App. Sci. Report No. E249.2.*

Chamieh, D.S., Acosta, A.J., Brennen, C.E., and Caughey, T.K. (1985). Experimental measurements of hydrodynamic radial forces and stiffness matrices for a centrifugal pump-impeller. *ASME J. Fluids Eng.*, **107**, No. 3, 307–315.

Childs, D.W. (1983a). Dynamic analysis of turbulent annular seals based on Hirs' lubrication equation. *ASME J. Lubr. Tech.*, **105**, 429–436.

Childs, D.W. (1983b). Finite length solutions for rotordynamic coefficients of turbulent annular seals. *ASME J. Lubr. Tech.*, **105**, 437–445.

Childs, D.W. (1987). Fluid structure interaction forces at pump-impeller-shroud surfaces for rotordynamic calculations. *ASME Symp. on Rotating Machinery Dynamics*, **2**, 581–593.

Childs, D.W. (1989). Fluid structure interaction forces at pump-impeller-shroud surfaces for rotordynamic calculations. *ASME J. Vibration, Acoustics, Stress and Reliability in Design*, **111**, 216–225.

Childs, D.W. and Dressman, J.B. (1982). Testing of turbulent seals for rotordynamic coefficients. *Proc. Workshop on Rotordynamic Instability Problems in High-Performance Turbomachinery, NASA Conf. Publ. 2250*, 157–171.

Childs, D.W. and Dressman, J.B. (1985). Convergent-tapered annular seals: analysis and testing for rotordynamic coefficients. *ASME J. Tribology*, **107**, 307–317.

Childs, D.W. and Kim, C.-H. (1985). Analysis and testing for rotordynamic coefficients of turbulent annular seals with different directionally homogeneous surface roughness treatment for rotor and stator elements. *ASME J. Tribology*, **107**, 296–306.

Childs, D.W. and Scharrer, J.K. (1986). Experimental rotordynamic coefficient results for teeth-on-rotor and teeth-on-stator labyrinth gas seals. *Proc. Adv. Earth-to-Orbit Propulsion Tech. Conf., NASA Conf. Publ. 2436*, 327–345.

Chivers, T.C. (1969). Cavitation in centrifugal pumps. *Proc. Inst. Mech. Eng.*, **184**, Part I, No. 2, 37–68.

Chu, S., Dong, R., and Katz, J. (1993). The noise characteristics within the volute of a centrifugal pump for different tongue geometries. *Proc. ASME Symp. on Flow Noise, Modelling, Measurement and Control.*

Colding-Jorgensen, J. (1979). The effect of fluid forces on rotor stability of centrifugal compressors and pumps. *Ph.D. Thesis, Tech. Univ. of Denmark.*

Constant, H. (1939). Performance of cascades of aerofoils. *Royal Aircraft Est. Note No. E3696 and ARC Rep. No. 4155.*

Cooper, P. (1967). Analysis of single- and two-phase flows in turbo-pump inducers. *ASME J. Eng. Power*, **89**, 577–588.

Csanady, G.T. (1964). *Theory of turbomachines.* McGraw-Hill, New York.

Cumpsty, N.A. (1977). Review—a critical review of turbomachinery noise. *ASME J. Fluids Eng.*, **99**, 278–293.

d'Agostino, L. and Brennen, C.E. (1983). On the acoustical dynamics of bubble clouds. *ASME Cavitation and Multiphase Flow Forum*, 72–75.

d'Agostino, L., Brennen, C.E., and Acosta, A.J. (1988). Linearized dynamics of two-dimensional bubbly and cavitating flows over slender surfaces. *J. Fluid Mech.*, **192**, 485–509.

d'Agostino, L. and Brennen, C.E. (1989). Linearized dynamics of spherical bubble clouds. *J. Fluid Mech.*, **199**, 155–176.

Dean, R.C. (1959). On the necessity of unsteady flow in fluid machines. *ASME J. Basic Eng.*, **81**, 24–28.

del Valle, J., Braisted, D.M., and Brennen, C.E. (1992). The effects of inlet flow modification on cavitating inducer performance. *ASME J. Turbomachinery*, **114**, 360–365.

De Siervi, F., Viguier, H.C., Greitzer, E.M., and Tan, C.S. (1982). Mechanisms of inlet-vortex formation. *J. Fluid Mech.*, **124**, 173–207.

Dixon, S.L. (1978). *Fluid mechanics, thermodynamics of turbomachinery.* Pergamon Press.

Domm, H. and Hergt, P. (1970). Radial forces on impeller of volute casing pumps. In *Flow Research on Blading* (ed: L.S. Dzung), Elsevier Publ. Co., 305–321.

Dowson, D. and Taylor, C.M. (1979). Cavitation in bearings. *Ann. Rev. Fluid Mech.*, **11**, 35–66.

Doyle, H.E. (1980). Field experiences with rotordynamic instability in high-performance turbomachinery. *Proc. First Workshop on Rotordynamic Instability Problems in High-Performance Turbomachinery, NASA Conf. Pub. 2133*, 3–13.

Dring, R.P., Joslyn, H.D., Hardin, L.W., and Wagner, J.H. (1982). Turbine rotor-stator interaction. *ASME J. Eng. for Power*, **104**, 729–742.

Duller, G.A. (1966). On the linear theory of cascades of supercavitating hydrofoils. *U.K. Nat. Eng. Lab. Rep. No. 218.*

Duncan, A.B. (1966). Vibrations in boiler feed pumps: a critical review of experimental and service experience. *Proc. Inst. Mech. Eng.*, **181**, 55–64.

Dussourd, J.L. (1968). An investigation of pulsations in the boiler feed system of a central power station. *ASME J. Basic Eng.*, **90**, 607–619.

Eck, B. (1973). *Fans.* Pergamon Press, London.

Eckardt, D. (1976). Detailed flow investigations with a high-speed centrifugal compressor impeller. *ASME J. Fluids Eng.*, **98**, 390–420.

Ehrich, F. and Childs, D. (1984). Self-excited vibration in high-performance turbomachinery. *Mech. Eng.*, May 1984, 66–79.

Ek, B. (1957). *Technische Strömungslehre.* Springer-Verlag.

Ek, M.C. (1978). Solution of the subsynchronous whirl problem in the high pressure hydrogen turbomachinery of the Space Shuttle Main Engine. *Proc. AIAA/SAE 14th Joint Propulsion Conf., Las Vegas, Nevada, Paper No. 78-1002.*

Emmons, H.W., Pearson, C.E., and Grant, H.P. (1955). Compressor surge and stall propagation. *Trans. ASME*, **79**, 455–469.

Emmons, H.W., Kronauer, R.E., and Rockett, J.A. (1959). A survey of stall propagation—experiment and theory. *ASME J. Basic Eng.*, **81**, 409–416.

Falvey, H.T. (1990). Cavitation in chutes and spillways. *US Bur. of Reclamation, Eng. Monograph No. 42.*

Fanelli, M. (1972). Further considerations on the dynamic behaviour of hydraulic turbomachinery. *Water Power*, June 1972, 208–222.

Ferguson, T.B. (1963). *The centrifugal compressor stage.* Butterworth, London.

Fischer, K. and Thoma, D. (1932). Investigation of the flow conditions in a centrifugal pump. *Trans. ASME, Hydraulics*, **54**, 141–155.

Fitzpatrick, H.M. and Strasberg, M. (1956). Hydrodynamic sources of sound. *Proc. First ONR Symp. on Naval Hydrodynamics*, 241–280.

France, D. (1986). Rotor instability in centrifugal pumps. *Shock and Vibr. Digest of Vibr. Inst.*, Jan.1986, 9–13.

Franklin, R.E. and McMillan, J. (1984). Noise generation in cavitating flows, the submerged jet. *ASME J. Fluids Eng.*, **106**, 336–341.

Franz, R. and Arndt, N. (1986). Measurements of hydrodynamic forces on the impeller of the HPOTP of the SSME. *Calif. Inst. of Tech., Div. Eng. and Appl. Sci., Report No. E249.*

Franz, R., Acosta, A.J., Brennen, C.E., and Caughey, T.K. (1989). The rotordynamic forces on a centrifugal pump impeller in the presence of cavitation. *Proc. ASME Symp. Pumping Machinery*, **FED-81**, 205–212.

Franz, R., Acosta, A.J., Brennen, C.E., and Caughey, T.K. (1990). The rotordynamic forces on a centrifugal pump impeller in the presence of cavitation. *ASME J. Fluids Eng.*, **112**, 264–271.

Fritz, R.J. (1970). The effects of an annular fluid on the vibrations of a long rotor. Part I—Theory and Part II—Test. *ASME J. Basic Eng.*, **92**, 923–937.

Fung, Y.C. (1955). *An introduction to the theory of aeroelasticity.* John Wiley and Sons.

Fujikawa, S. and Akamatsu, T. (1980). Effects of the non-equilibrium condensation of vapour on the pressure wave produced by the collapse of a bubble in a liquid. *J. Fluid Mech.*, **97**, 481–512.

Furuya, O. and Acosta, A.J. (1973). A note on the calculation of supercavitating hydrofoils with rounded noses. *ASME J. Fluids Eng.*, **95**, 222–228.

Furuya, O. (1974). Supercavitating linear cascades with rounded noses. *ASME J. Basic Eng., Series D*, **96**, 35–42.

Furuya, O. (1975). Exact supercavitating cascade theory. *ASME J. Fluids Eng.*, **97**, 419–429.

Furuya, O. (1985). An analytical model for prediction of two-phase (non-condensable) flow pump performance. *ASME J. Fluids Eng.*, **107**, 139–147.

Furuya, O. and Maekawa, S. (1985). An analytical model for prediction of two-phase flow pump performance — condensable flow case. *ASME Cavitation and Multiphase Flow Forum*, **FED-23**, 74–77.

Gallus, H.E. (1979). High speed blade-wake interactions. *von Karman Inst. for Fluid Mech. Lecture Series 1979-3*, **2**.

Gallus, H.E., Lambertz, J., and Wallmann, T. (1980). Blade-row interaction in an axial flow subsonic compressor stage. *ASME J. Eng. for Power*, **102**, 169–177.

Gates, E.M. and Acosta, A.J. (1978). Some effects of several free stream factors on cavitation inception on axisymmetric bodies. *Proc. 12th ONR Symp. on Naval Hydrodynamics*, 86–108.

Gavrilov, L.R. (1970). Free gas content of a liquid and acoustical techniques for its measurement. *Sov. Phys.—Acoustics*, **15**, No. 3, 285–295.

Gongwer, C. (1941). A theory of cavitation flow in centrifugal-pump impellers. *Trans. ASME*, **63**, 29–40.

Grabow, G. (1964). Radialdruck bei Kreiselpumpen. *Pumpen und Verdichter*, No. 2, 11–19.

Greitzer, E.M. (1976). Surge and rotating stall in axial flow compressors. Part I: Theoretical compression system model. Part II: Experimental results and comparison with theory. *ASME J. Eng. for Power*, **98**, 190–211.

Greitzer, E.M. (1981). The stability of pumping systems—the 1980 Freeman Scholar Lecture. *ASME J. Fluids Eng.*, **103**, 193–242.

Grist, E. (1974). NPSH requirements for avoidance of unacceptable cavitation erosion in centrifugal pumps. *Proc. I.Mech.E. Conf. on Cavitation*, 153–163.

Gross, L.A. (1973). An experimental investigation of two-phase liquid oxygen pumping. *NASA TN D-7451*.

Guinard, P., Fuller, T. and Acosta, A.J. (1953). Experimental study of axial flow pump cavitation. *Calif. Inst. of Tech. Hydro. Lab. Report, E-19.3*.

Guinzburg, A., Brennen, C.E., Acosta, A.J., and Caughey, T.K. (1990). Measurements of the rotordynamic shroud forces for centrifugal pumps. *Proc. ASME Turbomachinery Forum*, **FED-96**, 23–26.

Ham, N.D. (1968). Aerodynamic loading on a two-dimensional airfoil during dynamic stall. *AIAA J.*, **6**, 1927–1934.

Hartmann, M.J. and Soltis, R.F. (1960). Observation of cavitation in a low hub-tip ratio axial flow pump. *Proc. Gas Turbine Power and Hydraulic Conf., ASME Paper No. 60-HYD-14*.

Henderson and Tucker (1962). Performance investigation of some high speed pump inducers. *R.P.E. Tech. Note 214*. Referred to by Janigro and Ferrini (1973) but not located by the author.

Hennyey, Z. (1962). *Linear electric circuits*. Pergamon Press.

Hergt, P. and Benner, R. (1968). Visuelle Untersuchung der Strömung in Leitrad einer Radialpumpe. *Schweiz. Banztg.*, **86**, 716–720.

Hergt, P. and Krieger, P. (1969-70). Radial forces in centrifugal pumps with guide vanes. *Proc. Inst. Mech. Eng.*, **184**, Part 3N, 101–107.

Herrig, L.J., Emery, J.C., and Erwin, J.R. (1957). Systematic two-dimensional cascade tests of NACA 65-series compressor blades at low speeds. *NACA TN 3916*.

Hickling, R. and Plesset, M.S. (1964). Collapse and rebound of a spherical bubble in water. *Phys. Fluids*, **7**, 7–14.

Hirs, G.G. (1973). A bulk-flow theory for turbulence in lubricant films. *ASME J. of Lubr. Tech.*, April 1973, 137–146.

Hobson, D.E. and Marshall, A. (1979). Surge in centrifugal pumps. *Proc. 6th Conf. on Fluid Machinery, Budapest*, 475–483.

Hobbs, J.M., Laird, A., and Brunton, W.C. (1967). Laboratory evaluation of the vibratory cavitation erosion test. *Nat. Eng. Lab. (U.K.), Report No. 271.*

Holl, J.W. and Wislicenus, G.F. (1961). Scale effects on cavitation. *ASME J. Basic Eng.*, **83**, 385–398.

Holl, J.W. and Treaster, A.L. (1966). Cavitation hysteresis. *ASME J. Basic Eng.*, **88**, 199–212.

Holl, J.W. (1969). Limited cavitation. *Cavitation State of Knowledge*, ASME, 26–63.

Hori, Y. (1959). A theory of oil whip. *ASME J. Appl. Mech.*, June 1959, 189–198.

Horlock, J.M. (1968). Fluctuating lift forces on airfoils moving through transverse and chordwise gusts. *ASME J. Basic Eng.*, 494–500.

Horlock, J.H. (1973). *Axial flow compressors*. Robert E. Krieger Publ. Co., New York.

Horlock, J.H. and Lakshminarayana, B. (1973). Secondary flows: theory, experiment and application in turbomachinery aerodynamics. *Ann. Rev. Fluid Mech.*, **5**, 247–279.

Howard, J.H.G. and Osborne, C. (1977). A centrifugal compressor flow analysis employing a jet-wake passage model. *ASME J. Fluids Eng.*, **99**, 141–147.

Howell, A.R. (1942). The present basis of axial flow compressor design: Part I—Cascade theory and performance. *ARC R and M No. 2095.*

Hsu, C.C. (1972). On flow past a supercavitating cascade of cambered blades. *ASME J. Basic Eng., Series D*, **94**, 163–168.

Hydraulic Institute, New York. (1965). *Standards of the Hydraulic Institute* (11th edition).

Iino, T. and Kasai, K. (1985). An analysis of unsteady flow induced by interaction between a centrifugal impeller and a vaned diffuser (in Japanese). *Trans. Japan Soc. of Mech. Eng.*, **51**, No.471, 154–159.

Iversen, H.W., Rolling, R.E., and Carlson, J.J. (1960). Volute pressure distribution, radial force on the impeller and volute mixing losses of a radial flow centrifugal pump. *ASME J. Eng. for Power*, **82**, 136–144.

Jaeger, C. (1963). The theory of resonance in hydro-power systems, discussion of incidents and accidents occurring in pressure systems. *ASME J. Basic Eng.*, **85**, 631–640.

Jakobsen, J.K. (1964). On the mechanism of head breakdown in cavitating inducers. *ASME J. Basic Eng.*, **86**, 291–304.

Jakobsen, J.K. (1971). *Liquid rocket engine turbopumps*. NASA SP 8052.

Janigro, A. and Ferrini, F. (1973). Inducer pumps. In *Recent progress in pump research*, von Karman Inst. for Fluid Dynamics, Lecture Series 61.

Jansen, W. 1964. Rotating stall in a radial vaneless diffuser. *ASME J. Basic Eng.*, **86**, 750–758.

Japikse, D. (1984). *Turbomachinery diffuser design technology*. Concepts ETI, Inc., Norwich, VT.

Jery, B. and Franz, R. (1982). Stiffness matrices for the Rocketdyne diffuser volute. *Calif. Inst. of Tech., Div. Eng. and Appl. Sci., Report No. E249.1.*

Jery, B., Acosta, A.J., Brennen, C.E., and Caughey, T.K. (1985). Forces on centrifugal pump impellers. *Proc. Second Int. Pump Symp., Houston, Texas*, 21–32.

Johnson, V.E. and Hsieh, T. (1966). The influence of the trajectories of gas nuclei on cavitation inception. *Proc. 6th ONR Symp. on Naval Hydrodynamics*, 7-1.

Johnsson, C.A. (1969). Cavitation inception on headforms, further tests. *Proc. 12th Int. Towing Tank Conf., Rome*, 381–392.

Johnston, J.P. and Dean, R.C. (1966). Losses in vaneless diffusers of centrifugal compressors and pumps. *ASME J. Eng. for Power*, **88**, 49–62.

KSB. (1975). *KSB Centrifugal pump lexicon*. Klein, Schanzlin and Becker, Germany.

Kamijo, K., Shimura, T., and Watanabe, M. (1977). An experimental investigation of cavitating inducer instability. *ASME Paper 77-WA/FW-14.*

Kamijo, K., Shimura, T., and Watanabe, M. (1980). A visual observation cavitating inducer instability. *Nat. Aero. Lab. (Japan), Rept. NAL TR-598T.*

Kamijo, K., Yoshida, M., and Tsujimoto, Y. (1992). Hydraulic and mechanical performance of LE-7 LOX pump inducer. *Proc. 28th Joint Propulsion Conf., Paper AIAA-92-3133.*

Karyeaclis, M.P., Miskovish, R.S., and Brennen, C.E. (1989). Rotordynamic tests in cavitation of the SEP inducer. *Calif. Inst. of Tech., Div. of Eng. and Appl. Sci., Report E200.27.*

Katsanis, T. (1964). Use of arbitrary quasi-orthogonals for calculating flow distribution in the meridional plane of a turbomachine. *NASA TN D-2546.*

Katsanis, T. and McNally, W.D. (1977). Revised Fortran program for calculating velocities and streamlines on the hub-shroud midchannel stream surface of an axial-, radial-, or mixed-flow turbomachine or annular duct. *NASA TN D-8430 and D-8431.*

Katz, J., Gowing, S., O'Hern, T., and Acosta, A.J. (1984). A comparitive study between holographic and light-scattering techniques of microbubble detection. *Proc. IUTAM Symp. on Measuring Techniques in Gas-Liquid Two-Phase Flows*, Springer-Verlag, 41–66.

Keller, A.P. (1974). Investigations concerning scale effects of the inception of cavitation. *Proc. I.Mech.E. Conf. on Cavitation*, 109–117.

Keller, A.P. and Weitendorf, E.A. (1976). Influence of undissolved air content on cavitation phenomena at the propeller blades and on induced hull pressure amplitudes. *Proc. IAHR Symp. on two-phase flow and cavitation in power generation*, 65–76.

Kemp, N.H. and Ohashi, H. (1975). Forces on unstaggered airfoil cascades in unsteady in-phase motion with applications to harmonic oscillation. *Proc. Symp. on Unsteady Aerodynamics, Tuscon, Ariz.*, 793–826.

Kemp, N.H. and Sears, W.R. (1955). The unsteady forces due to viscous wakes in turbomachines. *J. Aero Sci.*, **22**, No.7, 478–483.

Kermeen, R.W. (1956). Water tunnel tests of NACA 4412 and Walchner Profile 7 hydrofoils in noncavitating and cavitating flows. *Calif. Inst. of Tech., Hydrodynamics Lab. Rep. 47-5.*

Kerrebrock, J.L. (1977). *Aircraft engines and gas turbines.* MIT Press.

Kimoto, H. (1987). An experimental evaluation of the effects of a water microjet and a shock wave by a local pressure sensor. *ASME Int. Symp. on Cavitation Res. Fac. and Techniques*, **FED-57**, 217–224.

Knapp, R.T., Daily, J.W., and Hammitt, F.G. (1970). *Cavitation.* McGraw-Hill, New York.

König, E. (1922). Potentialströmung durch Gitter. *Z. angew. Math. u. Mech.*, **2**, 422.

Lakshminarayana, B. (1972). Visualization study of flow in axial flow inducer. *ASME J. Basic Eng.*, **94**, 777–787.

Lakshminarayana, B. (1981). Analytical and experimental study of flow phenomena in noncavitating rocket pump inducers. *NASA Contractor Rep. 3471.*

Lauterborn, W. and Bolle, H. (1975). Experimental investigations of cavitation bubble collapse in the neighborhood of a solid boundary. *J. Fluid Mech.*, **72**, 391–399.

Lazarkiewicz, S. and Troskolanski, A.T. (1965). *Pompy wirowe. (Impeller pumps.)* Translated from Polish by D.K. Rutter. Publ. by Pergamon Press.

Lee, C.J.M. (1966). Written discussion in *Proc. Symp. on Pump Design, Testing and Operation, Natl. Eng. Lab., Scotland*, 114–115.

Lefcort, M.D. (1965). An investigation into unsteady blade forces in turbomachines. *ASME J. Eng. for Power*, **87**, 345–354.

Lenneman, E. and Howard, J.H.G. (1970). Unsteady flow phenomena in centrifugal impeller passages. *ASME J. Eng. for Power*, **92-1**, 65–72.

Lieblein, S., Schwenk, F.C., and Broderick, R.L. (1953). Diffusion factor for estimating losses and limiting blade loadings in axial-flow-compressor blade elements. *NACA RM E53D01.*

Lieblein, S. (1965). Experimental flow in two-dimensional cascades. *Aerodynamic design of axial flow compressors, NASA SP-36*, 183–226.

Lindgren, H. and Johnsson, C.A. (1966). Cavitation inception on headforms, ITTC comparative experiments. *Proc. 11th Int. Towing Tank Conf., Tokyo*, 219–232.

Lomakin, A.A. (1958). Calculation of the critical speed and the conditions to ensure dynamic stability of the rotors in high pressure hydraulic machines, taking account of the forces in the seals (in Russian). *Energomashinostroenie*, **14**, No. 4, 1–5.

Lorett, J.A. and Gopalakrishnan, S. (1983). Interaction between impeller and volute of pumps at off-design conditions. *Proc. ASME Symp. on Performance Characteristics of Hydraulic Turbines and Pumps*, **FED-6**, 135–140.

Lush, P.A. and Angell, B. (1984). Correlation of cavitation erosion and sound pressure level. *ASME J. Fluids Eng.*, **106**, 347–351.

McCroskey, W.J. (1977). Some current research in unsteady fluid dynamics—the 1976 Freeman Scholar Lecture. *ASME J. Fluids Eng.*, **99**, 8–38.

McNulty, P.J. and Pearsall, I.S. (1979). Cavitation inception in pumps. *ASME Int. Symp. on Cavitation Inception*, 163–170.

Makay, E. and Szamody, O. (1978). Survey of feed pump outages. *Electric Power Res. Inst. Rep. FP-754*.

Makay, E. (1980). Centrifugal pump hydraulic instability. *Electric Power Res. Inst. Rep. EPRI CS-1445*.

Mansell, C.J. (1974). Impeller cavitation damage on a pump operating below its rated discharge. *Proc. of Conf. on Cavitation, Inst. of Mech. Eng.*, 185–191.

Marscher, W.D. (1988). Subsynchronous vibration in boiler feed pumps due to stable response to hydraulic forces on at part-load. *Proc. I.Mech.E. Conf. on Part Load Pumping, London*, 167–175.

Martin, C.S. (1978). Waterhammer in steam generators and feedwater piping. *Paper presented at Steam Generator Waterhammer Workshop, EPRI, Palo Alto, Cal.*

Martin, M. (1962). Unsteady lift and moment on fully cavitating hydrofoils at zero cavitation number. *J. Ship Res.*, **6**, No.1, 15–25.

Medwin, H. (1977). In situ acoustic measurements of microbubbles at sea. *J. Geophys. Res.*, **82**, No.6, 921–976.

Meyer, R.X. (1958). The effect of wakes on the transient pressure and velocity distributions in turbomachines. *Trans. ASME*, **80**, 1544.

Mikolajczak, A.A., Arnoldi, R.A., Snyder, L.E., and Stargardter, H. (1975). Advances in fan and compressor blade flutter analysis and predictions. *J. Aircraft*, **12**, No.4, 325–332.

Miller, C.D. and Gross, L.A. (1967). A performance investigation of an eight-inch hubless pump inducer in water and liquid nitrogen. *NASA TN D-3807*.

Mimura, Y. (1958). The flow with wake past an oblique plate. *J. Phys. Soc. Japan*, **13**, 1048–1055.

Miskovish, R.S. and Brennen, C.E. (1992). Some unsteady fluid forces measured on pump impellers. *ASME J. Fluids Eng.*, **114**, 632–637.

Miyagawa, K., Mutaguchi, K., Kanki, H., Iwasaki, Y., Sakamoto, A., Fujiki, S., Terasaki, A., and Furuya, S. (1992). An experimental investigation of fluid exciting force on a high head pump-turbine runner. *Proc. 4th Int. Symp. on Transport Phenomena and Dynamics of Rotating Machinery*, **B**, 133–142.

Moore, R.D. and Meng, P.R. (1970a). Thermodynamic effects of cavitation of an 80.6° helical inducer operated in hydrogen. *NASA TN D-5614*.

Moore, R.D. and Meng, P.R. (1970b). Effect of blade leading edge thickness on cavitation performance of 80.6° helical inducers in hydrogen. *NASA TN D5855*.

Murai, H. (1968). Observations of cavitation and flow patterns in an axial flow pump at low flow rates (in Japanese). *Mem. Inst. High Speed Mech., Tohoku Univ.*, **24**, No.246, 315–333.

Murakami, M. and Minemura, K. (1977). Flow of air bubbles in centrifugal impellers and its effect on the pump performance. *Proc. 6th Australasian Hydraulics and Fluid Mechanics Conf.*, **1**, 382–385.

Murakami, M. and Minemura, K. (1978). Effects of entrained air on the performance of a horizontal axial-flow pump. In *Polyphase Flow in Turbomachinery* (eds. C.E. Brennen, P. Cooper and P.W. Runstadler, Jr.), ASME, 171–184.

Myles, D.J. (1966). A design method for mixed flow pumps and fans. *Proc. Symp. on Pump Design, Testing and Operation, Nat. Eng. Lab., Scotland*, 167–176.

NASA. (1970). Prevention of coupled structure-propulsion instability. *NASA SP-8055.*

Natanzon, M.S., Bl'tsev, N.E., Bazhanov, V.V., and Leydervarger, M.R. (1974). Experimental investigation of cavitation-induced oscillations of helical inducers. *Fluid Mech., Soviet Res.*, **3**, No.1, 38–45.

Naude, C.F. and Ellis, A.T. (1961). On the mechanism of cavitation damage by non-hemispherical cavities in contact with a solid boundary. *ASME. J. Basic Eng.*, **83**, 648–656.

Newkirk, B.L. and Taylor, H.D. (1925). Shaft whipping due to oil action in journal bearing. *General Electric Review, Aug. 1925*, 559–568.

Ng, S.L. and Brennen, C.E. (1978). Experiments on the dynamic behavior of cavitating pumps. *ASME J. Fluids Eng.*, **100**, No. 2, 166–176.

Nordmann, R. and Massmann, H. (1984). Identification of dynamic coefficients of annular turbulent seals. *Proc. Workshop on Rotordynamic Instability Problems in High-Performance Turbomachinery, NASA Conf. Publ. 2338*, 295–311.

Numachi, F. (1961). Cavitation tests on hydrofoils designed for accelerating flow cascade: Report 1. *ASME J. Basic Eng.*, **83**, Series D, 637–647.

Numachi, F. (1964). Cavitation tests on hydrofoils designed for accelerating flow cascade: Report 3. *ASME J. Basic Eng.*, **86**, Series D, 543–555.

Ohashi, H. (1968). Analytical and experimental study of dynamic characteristics of turbo-pumps. *NASA TN D-4298.*

Ohashi, H. and Shoji, H. (1984a). Theoretical study of fluid forces on whirling centrifugal impeller (in Japanese). *Trans. JSME*, **50**, No. 458 B, 2518–2523.

Ohashi, H. and Shoji, H. (1984b). Lateral fluid forces acting on a whirling centrifugal impeller in vaneless and vaned diffuser. *Proc. Workshop on Rotordynamic Instability Problems in High Performance Turbomachinery, NASA Conf. Publ. 2338*, 109–122.

Okamura, T. and Miyashiro, H. (1978). Cavitation in centrifugal pumps operating at low capacities. *ASME Symp. on Polyphase Flow in Turbomachinery*, 243–252.

Omta, R. (1987). Oscillations of a cloud of bubbles of small and not so small amplitude. *J. Acoust. Soc. Amer.*, **82**, 1018–1033.

Ooi, K.K. (1985). Scale effects on cavitation inception in submerged water jets: a new look. *J. Fluid Mech.*, **151**, 367–390.

Oshima, M. and Kawaguchi, K. (1963). Experimental study of axial and mixed flow pumps. *Proc. IAHR Symp. on Cavitation and Hydraulic Machinery, Sendai, Japan*, 397–416.

Parkin, B.R. (1952). Scale effects in cavitating flow. *Ph.D. Thesis, Calif. Inst. of Tech., Pasadena.*

Parkin, B.R. (1958). Experiments on circular-arc and flat plate hydrofoils. *J. Ship Res.*, **1**, 34–56.

Parkin, B.R. (1962). Numerical data on hydrofoil reponse to non-steady motions at zero cavitation number. *J. Ship Res.*, **6**, 40–42.

Patel, B.R. and Runstadler, P.W., Jr. (1978). Investigations into the two-phase flow behavior of centrifugal pumps. In *Polyphase Flow in Turbomachinery* (eds: C.E. Brennen, P. Cooper and P.W. Runstadler, Jr.), ASME, 79–100.

Paynter, H.M. (1961). *Analysis and design of engineering systems*. MIT Press.

Pearsall, I.S. (1963). Supercavitation for pumps and turbines. *Engineering (GB)*, **196(5081)**, 309–311.

Pearsall, I.S. (1966-67). Acoustic detection of cavitation. *Proc. Inst. Mech. Eng.*, **181**, No. 3A.

Pearsall, I.S. (1972). *Cavitation*. Mills & Boon Ltd., London.

Pearsall, I.S. (1978). Off-design performance of pumps. *von Karman Inst. for Fluid Dynamics, Lecture Series 1978-3.*

Peck, J.F. (1966). Written discussion in *Proc. Symp. on Pump Design, Testing and Operation, Nat. Eng. Lab., Scotland,* 256–273.

Peterson, F.B., Danel, F., Keller, A.P., and Lecoffre, Y. (1975). Determination of bubble and particulate spectra and number density in a water tunnel with three optical techniques. *Proc. 14th Int. Towing Tank Conf., Ottawa,* **2**, 27–52.

Pfleiderer, C. (1932). *Die Kreiselpumpen.* Julius Springer, Berlin.

Pinkus, O. and Sternlicht, B. (1961). *Theory of hydrodynamic lubrication.* McGraw-Hill, New York.

Pipes, L.A. (1940). The matrix theory for four terminal networks. *Phil. Mag.,* **30**, 370.

Pipes, L.A. (1963). *Matrix methods for engineering.* Prentice-Hall, Inc., NJ.

Platzer, M. F. (1978). Unsteady flows in turbomachines—a review of current developments. *AGARD Rept. CP-227.*

Plesset, M.S. (1949). The dynamics of cavitation bubbles. *Trans. ASME, J. Appl. Mech.,* **16**, 228–231.

Plesset, M.S. and Chapman, R.B. (1971). Collapse of an initially spherical vapor cavity in the neighborhood of a solid boundary. *J. Fluid Mech.,* **47**, 283–290.

Plesset, M.S. and Prosperetti, A. (1977). Bubble dynamics and cavitation. *Ann. Rev. Fluid Mech.,* **9**, 145–185.

Pollman, E., Schwerdtfeger, H., and Termuehlen, H. (1978). Flow excited vibrations in high pressure turbines (steam whirl). *ASME J. Eng. for Power,* **100**, 219–228.

Rains, D.A. (1954). Tip clearance flows in axial flow compressors and pumps. *Calif. Inst. of Tech. Hydro. and Mech. Eng. Lab. Report, No. 5.*

Rayleigh, Lord. (1917). On the pressure developed in a liquid during the collapse of a spherical cavity. *Phil. Mag.,* **34**, 94–98.

Rohatgi, U.S. (1978). Pump model for two-phase transient flow. In *Polyphase Flow in Turbomachinery* (eds: C.E. Brennen, P. Cooper and P.W. Runstadler, Jr.), ASME, 101–120.

Rosenmann, W. (1965). Experimental investigations of hydrodynamically induced shaft forces with a three bladed inducer. *Proc. ASME Symp. on Cavitation in Fluid Machinery,* 172–195.

Roudebush, W.H. (1965). Potential flow in two-dimensional cascades. *Aerodynamic design of axial flow compressors, NASA SP-36,* 101–149.

Roudebush, W.H. and Lieblein, S. (1965). Viscous flow in two-dimensional cascades. *Aerodynamic design of axial flow compressors, NASA SP-36,* 151–181.

Rubin, S. (1966). Longitudinal instability of liquid rockets due to propulsion feedback (Pogo). *J. Spacecraft and Rockets,* **3**, No.8, 1188–1195.

Ruggeri, R.S. and Moore, R.D. (1969). Method for prediction of pump cavitation performance for various liquids, liquid temperatures, and rotative speeds. *NASA TN D-5292.*

Sabersky, R.H., Acosta, A.J. and Hauptmann, E.G. (1989). *Fluid flow (3rd edition), Chapters 12 and 13.* Macmillan Publ. Co.

Sack, L.E. and Nottage, H.B. (1965). System oscillations associated with cavitating inducers. *ASME J. Basic Eng.,* **87**, 917–924.

Safwat, H.H. and van der Polder, J. (1973). Experimental and analytic data correlation study of water column separation. *ASME J. Fluids Eng.,* **95**, 91–97.

Salemann, V. (1959). Cavitation and NPSH requirements of various liquids. *ASME J. Basic Eng.,* **81**, 167–180.

Samoylovich, G.S. (1962). On the calculation of the unsteady flow around an array of arbitrary profiles vibrating with arbitrary phase shift. *Prikladnaya Matematika i Mekhanika,* No.4.

Sano, M. (1983). Pressure pulsations in turbo-pump piping systems. Experiments on the natural frequencies of the liquid columns in centrifugal pump piping systems. *Bull. JSME,* **26**, No.222, 2129–2135.

Schoeneberger, W. (1965). Cavitation tests in radial pump impellers. *Ph.D. Thesis, Tech. Univ. Darmstadt.*

Schorr, B. and Reddy, K.C. (1971). Inviscid flow through cascades in oscillatory and distorted flow. *AIAA J.*, **9**, 2043–2050.

Shima, A., Takayama, K., Tomita, Y., and Muira, N. (1981). An experimental study on effects of a solid wall on the motion of bubbles and shock waves in bubble collapse. *Acustica*, **48**, 293–301.

Shoji, H. and Ohashi, H. (1980). Fluid forces on rotating centrifugal impeller with whirling motion. *Proc. First Workshop on Rotordynamic Instability Problems in High-Performance Turbomachinery, NASA Conf. Pub. 2133*, 317–328.

Silberman, E. (1959). Experimental studies of supercavitating flow about simple two-dimensional bodies in a jet. *J. Fluid Mech.*, **5**, 337–354.

Silberman, E. and Song, C.S. (1961). Instability of ventilated cavities. *J. Ship Res.*, **5**, 13–33.

Sisto, F. (1953). Stall-flutter in cascades. *J. Aero. Sci.*, **20**, 598–604.

Sisto, F. (1967). Linearized theory of non-stationary cascades at fully stalled or supercavitating conditions. *Zeitschrift fur Angewandte Mathematik und Mechanik*, **8**, 531–542.

Sisto, F. (1977). A review of the fluid mechanics of aeroelasticity in turbomachines. *ASME J. Fluids Eng.*, **99**, 40–44.

Sloteman, D.P., Cooper, P., and Dussourd, J.L. (1984). Control of backflow at the inlets of centrifugal pumps and inducers. *Proc. Int. Pump Symp., Texas A&M Univ.*, 9–22.

Sloteman, D.P., Cooper, P., and Graf, E. (1991). Design of high-energy pump impellers to avoid cavitation instabilities and damage. *Proc. EPRI Power Plant Symp., Tampa, Fl., June 1991*.

Song, C.S. (1962). Pulsation of ventilated cavities. *J. Ship Res.*, **5**, 8–20.

Soyama, H., Kato, H., and Oba, R. (1992). Cavitation observations of severely erosive vortex cavitation arising in a centrifugal pump. *Proc. Third I.Mech.E. Int. Conf. on Cavitation*, 103–110.

Sparks, C.R. and Wachel, J.C. (1976). Pulsations in liquid pumps and piping systems. *Proc. 5th Turbomachinery Symp.*, 55–61.

Spraker, W.A. (1965). The effect of fluid properties on cavitation in centrifugal pumps. *ASME J. Eng. Power*, **87**, 309–318.

Stahl, H.A. and Stepanoff, A.J. (1956). Thermodynamic aspects of cavitation in centrifugal pumps. *Trans. ASME*, **78**, 1691–1693.

Stanitz, J.D. (1952). Some theoretical aerodynamic investigations of impellers in radial- and mixed- flow centrifugal compressors. *Trans. ASME*, **74**, 473–497.

Stepanoff, A.J. (1957). *Centrifugal and axial flow pumps*. John Wiley and Sons, Inc.

Stepanoff, A.J. (1961). Cavitation in centrifugal pumps with liquids other than water. *ASME J. Eng. Power*, **83**, 79–90.

Stepanoff, A.J. (1964). Cavitation properties of liquids. *ASME J. Eng. Power*, **86**, 195–200.

Stockman, N.O. and Kramer, J.L. (1963). Method for design of pump impellers using a high-speed digital computer. *NASA TN D-1562*.

Stodola, A. (1927). *Steam and gas turbines. Volumes I and II*. McGraw-Hill, New York.

Strecker, F. and Feldtkeller, R. (1929). Grundlagen der Theorie des allgemeinen Vierpols. *Elektrische Nachrichtentechnik*, **6**, 93.

Streeter, V.L. and Wylie, E.B. (1967). *Hydraulic transients*. McGraw-Hill.

Streeter, V.L. and Wylie, E.B. (1974). Waterhammer and surge control. *Ann. Rev. Fluid Mech.*, **6**, 57–73.

Stripling, L.B. and Acosta, A.J. (1962). Cavitation in turbopumps - Part I. *ASME J. Basic Eng.*, **84**, 326–338.

Stripling, L.B. (1962). Cavitation in turbopumps - Part II. *ASME J. Basic Eng.*, **84**, 339–350.

Strub, R.A. (1963). Pressure fluctuations and fatigue stresses in storage pumps and pump-turbines. *ASME Paper No. 63-AHGT-11*.

Sturge, D.P. and Cumpsty, N.A. (1975). Two-dimensional method for calculating separated flow in a centrifugal impeller. *ASME J. Fluids Eng.*, **97**, 581–579.

Sutherland, C.D. and Cohen, H. (1958). Finite cavity cascade flow. *Proc. 3rd U.S. Nat. Cong. of Appl. Math.*, 837–845.

Tanahashi, T. and Kasahara, E. (1969). Analysis of water hammer with water column separation. *Bull. JSME*, **12**, No.50, 206–214.

Thiruvengadam, A. (1967). The concept of erosion strength. Erosion by cavitation or impingement. *ASTM STP 408, Am. Soc. Testing Mats.*, 22.

Thiruvengadam, A. (1974). Handbook of cavitation erosion. *Tech. Rep. 7301-1, Hydronautics, Inc., Laurel, Md.*

Thompson, W.E. (1978). Fluid dynamic excitation of centrifugal compressor rotor vibrations. *ASME J. Fluids Eng.*, **100**, No. 1, 73–78.

Tsujimoto, Y., Imaichi, K., Tomohiro, T., and Gatoo, M. (1986). A two-dimensional analysis of unsteady torque on mixed flow impellers. *ASME J. Fluids Eng.*, **108**, No. 1, 26–33.

Tsujimoto, Y., Acosta, A.J., and Brennen, C.E. (1988). Theoretical study of fluid forces on a centrifugal impeller rotating and whirling in a volute. *ASME J. Vibration, Acoustics, Stress and Reliability in Design*, **110**, 263–269.

Tsujimoto, Y., Kamijo, K., and Yoshida, Y. (1992). A theoretical analysis of rotating cavitation in inducers. *ASME Cavitation and Multiphase Flow Forum*, **FED-135**, 159–166.

Tsukamoto, H. and Ohashi, H. (1982). Transient characteristics of centrifugal turbomachines. *ASME J. Fluids Eng.*, **104**, No. 1, 6–14.

Tulin, M.P. (1953). Steady two-dimensional cavity flows about slender bodies. *David Taylor Model Basin Rep. 834.*

Tulin, M.P. (1964). Supercavitating flows - small perturbation theory. *J. Ship Res.*, **7**, No. 3, 16–37.

Tyler, J.M. and Sofrin, T.G. (1962). Axial compressor noise studies. *Soc. Automotive Eng.*, **70**, 309–332.

van der Braembussche, R. (1982). Rotating stall in vaneless diffusers of centrifugal compressors. *von Karman Inst. for Fluid Dyn., Technical Note 145.*

Vaage, R.D., Fidler, L.E., and Zehnle, R.A. (1972). Investigation of characteristics of feed system instabilities. *Final Rept. MCR-72-107, Martin Marietta Corp., Denver, Col.*

Vance, J.M. (1988). *Rotordynamics of turbomachinery.* John Wiley and Sons, New York.

Verdon, J.M. (1985). Linearized unsteady aerodynamic theory. *United Technologies Research Center Report R85-151774-1.*

Wachter, J. and Benckert, H. (1980). Flow induced spring coefficients of labyrinth seals for application in rotordynamics. *Proc. Workshop on Rotordynamic Instability problems in High Performance Turbomachinery, NASA Conf. Publ. 2133*, 189–212.

Wade, R.B. and Acosta, A.J. (1966). Experimental observations on the flow past a plano-convex hydrofoil. *ASME J. Basic Eng.*, **88**, 273–283.

Wade, R.B. (1967). Linearized theory of a partially cavitating cascade of flat plate hydrofoils. *Appl. Sci. Res.*, **17**, 169–188.

Wade, R.B. and Acosta, A.J. (1967). Investigation of cavitating cascades. *ASME J. Basic Eng., Series D*, **89**, 693–706.

Warnock, J.E. (1945). Experiences of the Bureau of Reclamation. *Proc. Amer. Soc. Civil Eng.*, **71**, No.7, 1041–1056.

Weyler, M.E., Streeter, V.L., and Larsen, P.S. (1971). An investigation of the effect of cavitation bubbles on the momentum loss in transient pipe flow. *ASME J. Basic Eng.*, **93**, 1–10.

Whitehead, D. (1960). Force and moment coefficients for vibrating airfoils in cascade. *ARC R&M 3254, London.*

Wiesner, F.J. (1967). A review of slip factors for centrifugal impellers. *ASME J. Eng. for Power*, **89**, 558–576.

Wiggert, D.C. and Sundquist, M.J. (1979). The effect of gaseous cavitation on fluid transients. *ASME J. Fluids Eng.*, **101**, 79–86.

Wijdieks, J. (1965). Greep op het ongrijpbare—II. Hydraulische aspecten bij het ontwerpen van pompinstallaties. *Delft Hydraulics Laboratory Publ. 43.*

Wislicenus, G.F. (1947). *Fluid mechanics of turbomachinery.* McGraw-Hill, New York.

Wood, G.M. (1963). Visual cavitation studies of mixed flow pump impellers. *ASME J. Basic Eng.*, Mar. 1963, 17–28.

Woods, L.C. (1955). On unsteady flow through a cascade of airfoils. *Proc. Roy. Soc. A*, **228**, 50–65.

Woods, L.C. (1957). Aerodynamic forces on an oscillating aerofoil fitted with a spoiler. *Proc. Roy. Soc. A*, **239**, 328–337.

Woods, L.C. (1961). *The theory of subsonic plane flow.* Cambridge Univ. Press.

Woods, L.C. and Buxton, G.H.L. (1966). The theory of cascade of cavitating hydrofoils. *Quart. J. Mech. Appl. Math.*, **19**, 387–402.

Worster, R.C. (1963). The flow in volutes and its effect on centrifugal pump performance. *Proc. Inst. of Mech. Eng.*, **177**, No. 31, 843–875.

Wu, T.Y. (1956). A free streamline theory for two-dimensional fully cavitated hydrofoils. *J. Math. Phys.*, **35**, 236–265.

Wu, T.Y. (1962). A wake model for free streamline flow theory, Part 1. Fully and partially developed wake flows and cavity flows past an oblique flat plate. *J. Fluid Mech.*, **13**, 161–181.

Wu, T.Y. and Wang, D.P. (1964). A wake model for free streamline flow theory, Part 2. Cavity flows past obstacles of arbitrary profile. *J. Fluid Mech.*, **18**, 65–93.

Wu, T.Y. (1972). Cavity and wake flows. *Ann. Rev. Fluid Mech.*, **4**, 243–284.

Yamada, Y. (1962). Resistance of flow through an annulus with an inner rotating cylinder. *Bull. JSME*, **5**, No. 18, 302–310.

Yamamoto, K. (1991). Instability in a cavitating centrifugal pump. *JSME Int. J., Ser. II*, **34**, 9–17.

Yoshida, Y., Murakami, Y., Tsurusaki, T., and Tsujimoto, Y. (1991). Rotating stalls in centrifugal impeller/vaned diffuser systems. *Proc. First ASME/JSME Joint Fluids Eng. Conf.*, **FED-107**, 125–130.

Young, W.E., Murphy, R., and Reddecliff, J.M. (1972). Study of cavitating inducer instabilities. *Pratt and Whitney Aircraft, Florida Research and Development Center, Rept. PWA FR-5131.*

Index

Printed in the United States
By Bookmasters